Operations and Maintenance Manual for Energy Management

Operations and Maintenance Manual for Energy Management

James E. Piper

SHARPE PROFESSIONAL
An imprint of M.E. Sharpe, INC.

Library of Congress Cataloging-in-Publication Data

Piper, James E.
Operations and maintenance manual for energy management / James E. Piper.
p.   cm.
Includes index.
ISBN 0-7656-0050-1 (hc. : alk. paper)
1. Buildings—Energy conservation—Handbooks, manuals, etc.
2. Buildings—Energy consumption—Handbooks, manuals, etc.
3. Buildings—Mechanical equipment—Maintenance and repair—
Handbooks, manuals, etc.   I. Title.
TJ163.5.B84P57   1998
658.2—dc21   98-47335
CIP

Printed in the United States of America

The paper used in this publication meets the minimum requirements of
American National Standard for Information Sciences—
Permanence of Paper for Printed Library Materials,
ANSI Z 39.48-1984.

MV (c)   10   9   8   7   6   5   4   3   2   1

About the Author

James E. Piper is an engineer with more than 25 years of experience in the fields of energy management and facilities maintenance. He is a licensed professional engineer who has developed and implemented both energy and maintenance management programs for facilities ranging in size from a few thousand square feet to ones with more than ten million square feet. He is currently working as a facility consultant in Bowie, Maryland.

The author received both a Bachelor's degree and a Master of Science degree in Mechanical Engineering from the University of Akron, and a Ph.D. in Educational Administration from the University of Maryland. Dr. Piper has published more than 300 articles on a wide range of facilities management topics, including the *Handbook of Facility Management.*

What This Manual Will Do for You

Operations and Maintenance Manual for Energy Management is a complete reference that you can use to evaluate and improve your maintenance and energy management operations. Every day, you are faced with having to make decisions on how best to use available resources. There never is enough money to do all that you want or need to. As a result, maintenance managers and energy conservation program managers often find themselves competing for the same resources. What is needed is a guide to show you how both maintenance operations and energy management can be enhanced at the same time.

This book is written for those who are responsible for managing energy use in their facilities. It is designed to serve as a practical guide to energy conservation through sound operation and maintenance practices. It provides detailed, practical information on how to improve the energy efficiency of your facility without having to invest thousands or millions of dollars in energy conservation projects. It provides straight-forward information on operation and maintenance tasks that you can implement to reduce energy use. Examples of tasks and their economic benefits are presented throughout the book.

Here are just a few of the benefits that you will gain from this book:

- ☐ an understanding of the relationship between maintenance practices and energy conservation;
- ☐ a method to evaluate the energy performance of your facilities;
- ☐ maintenance activities that will help to reduce building chiller energy requirements;
- ☐ boiler operation and maintenance activities to minimize energy use and promote equipment life;
- ☐ HVAC operation and maintenance activities that improve system performance and reduce energy requirements;
- ☐ lighting system maintenance practices that increase light output while reducing energy requirements;
- ☐ operation and maintenance practices that increase the life of building exterior components while reducing energy requirements;
- ☐ how to establish a comprehensive maintenance program that helps to ensure that energy use is reduced and remains low throughout the life of the systems.

Today, even though the chances of a new oil embargo are slight, energy conservation continues to play an important role in managing the operation of facilities. To many facility managers, energy conservation means making significant investments in capital improvements: installation of high-efficiency chillers and boilers, building energy management systems, thermal storage systems, or high thermal efficiency windows. Projects such as these typically cost tens or hundreds of thousands of dollars to implement, cause significant disruptions to the day-to-day operation of facilities during construction, and take years to recover the investment in the form of energy savings.

While there is no doubt that energy conservation projects, if properly implemented, will reduce energy use, energy conservation projects are not sufficient on their own to manage energy use. Comprehensive energy management also requires that building systems and components be operated and maintained at their best possible operating efficiency. Without proper maintenance, those new, energy conserving systems being installed today will rapidly deteriorate in both performance and efficiency.

Energy conservation and maintenance are so interconnected that it is impossible to separate the activities that promote energy conservation from those that promote good maintenance. Energy management requires sound maintenance, and sound maintenance promotes energy conservation. Sound maintenance is so effective at conserving energy that the highest rates of return for energy conservation investments are nearly always associated with maintenance activities. Typical payback periods are measured in weeks and months, not years. One would be hard-pressed to find energy projects offering similar rates of return.

If sound maintenance practices are so effective at promoting energy conservation, why are they so often overlooked in energy management programs today? The problem is one of perception. To those in facility management who are not responsible for maintenance activities, maintenance is typically viewed as a necessary evil. An enormous amount of money is spent maintaining what you already have—money that then is unavailable for investment in programs that provide a return for the organization. Many have a hard time understanding why money must be spent on systems that are operating perfectly well. "If it ain't broke, don't fix it." Perhaps this is a major reason why traditionally, maintenance has been one of the first areas to be cut during tight fiscal times.

For those who are responsible for maintenance of facilities, the challenge is to reverse this belief. The challenge is even more difficult for those who are responsible for energy management. Those who control the budget must be convinced that financing maintenance is really an investment—an investment that provides rates of return that exceed all other investments made by the organization.

HOW TO USE THIS BOOK

This book is divided into five sections: The first section presents an overview of how maintenance managers have approached energy management in the past, and why energy management is more important today. The second examines building mechanical systems, identifying operation and maintenance activities that can be implemented to reduce energy use. The third section examines building electrical systems, identifying operation and maintenance practices that can be implemented to improve the energy efficiency of electrical systems ranging from lighting to motors. The fourth section examines the building envelope. Specific practices are suggested to reduce the impact of aging and wear and tear on the energy efficiency of components of the building envelope. Finally, the fifth section shows how to promote and establish a comprehensive maintenance program designed to increase equipment life and performance while minimizing energy requirements.

Figures and worksheets, along with step-by-step instructions and *Rules of Thumb* are provided throughout this handy reference. Some have been developed from recognized maintenance and energy authorities. Others have been developed exclusively for this book and cannot be found in any other source. All will help you in establishing an energy maintenance program that will reduce the cost of energy, decrease the frequency of equipment breakdowns, improve the reliability of your energy using systems, and reduce the total cost of maintenance in your facility.

Table of Contents

SECTION 1. ENERGY MAINTENANCE IN FACILITIES

CHAPTER 1. THE ENERGY–MAINTENANCE CONNECTION 3

A Traditional View Toward Maintenance 4
The Focus on Energy Projects 6
Lack of Glamour 7
The Energy Management Role of Maintenance 8
Promoting Maintenance as Good Energy Policy 9

CHAPTER 2. HOW ENERGY MANAGEMENT IS TRADITIONALLY
PERFORMED BY ORGANIZATIONS: A FIVE-PHASE PROCESS 11

The Sudden Need for Energy Management 12
    Phase I: Quick Fixes 13
    Phase II: Energy Projects 14
    Phase III: Comprehensive Energy Management 15
    Phase IV: Declining Interest 16
    Phase V: Coming Full-Circle 16

CHAPTER 3. KEY INCENTIVES FOR REDUCING ENERGY
REQUIREMENTS IN FACILITIES 19

Economic Incentives for Energy Management 19
Supply Incentives for Energy Management 20
    Domestic Oil Production 21
    Increasing Dependence on Imports 22
    Natural Gas Supplies 22
    Electrical Generating Capacity 23
    The Impact of Supply Shortfalls 23
Environmental Incentives for Energy Management 23
Where We Are Headed 24

CHAPTER 4. HOW TO EVALUATE BUILDING ENERGY
PERFORMANCE 26

How to Measure the Performance of Your Energy Budget 27
Factors That Influence Energy Use and Performance 28

People Factors 29
Building Type Factors 29
Occupancy Factors 30
Climate Factors 30
Age Factors 33
Construction Factors 33
Other Factors 37
How to Effectively Use Energy Performance Data 37

CHAPTER 5. HOW TO ANALYZE UTILITY BILLS IN ORDER
TO REDUCE ENERGY COSTS 40

Electrical Bill Components 41
kWh Charges 41
Demand Charges 42
Electrical Demand Ratchet Clauses 42
Power Factor Charges 42
Miscellaneous Charges 43
Electrical Rate Structures 43
Additional Rate Factors 45
Power Factor 45
High Voltage Service 45
Curtailable Service 46
Natural Gas Rate Structures 46
Using the Information 47

SECTION 2. MECHANICAL SYSTEMS OPERATION AND
MAINTENANCE PRACTICES THAT WILL REDUCE ENERGY USE

CHAPTER 6. CHILLER SYSTEMS 51

Chiller System Energy Maintenance Considerations 51
Four Types of Building Chillers and Their Energy Requirements 52
Centrifugal Chillers 53
Reciprocating Chillers 53
Rotary Chillers 54
Absorption Chillers 54
How to Maintain Building Chillers for Energy Efficiency 55
How Fouling Works in Chiller Systems 57
How to Reduce Fouling 59
How Corrosion Works in Chiller Systems 60
How to Reduce Corrosion 61
How Scale Works in Chiller Systems 61

How to Reduce Scale 62
How to Set Up Chiller Maintenance Tasks 62
Chiller Leaks 62
Tube Testing 65
Refrigerant Testing 66
Oil Testing 66
Absorption Chiller Maintenance Considerations 66
Maintaining the Vacuum 69
Controlling Corrosion 69
Chiller Operations for Energy Management 70
Chiller Load Allocation 70
Condenser Water Temperature Reset 71
Chilled Water Reset 71
Indirect Free Cooling 72
Cooling Tower Maintenance Considerations 72
Tower Maintenance Requirements 73
Fill Material 73
Drift Eliminators 74
Water Distribution System 74
Fans 75
Tower Structures 75
Tower Inspection 76
Implementing the Energy Maintenance Program 79
Summary 82

CHAPTER 7. BOILER SYSTEMS 83

Boiler System Energy Maintenance Considerations 83
Three Methods of Rating Boiler Efficiency 84
Using the Boiler Operating Log 86
Boiler Water Treatment 88
How to Control Boiler Blowdown 90
How to Control Excess Combustion Air 93
Burner Maintenance 93
Boiler Preventative Maintenance Activities 96
Steam Distribution System Components and How to Maintain Them 100
Steam Piping 100
Steam Condensate Systems 102
Steam Traps 103
Steam Trap Maintenance Program 105
Two Operating Practices for Energy Conservation 107
Load Management 107
Lowering Operating Pressure 108
Summary 108

CHAPTER 8. AIR HANDLING SYSTEMS   109

Air Handling System Energy Maintenance Considerations   109
Elements of the Air Handling System and Their Maintenance
    Requirements   111
       Fans   111
       Ventilation Rates and Indoor Air Quality   112
       Heating and Cooling Coils   116
       Filters   116
       Controls   118
       Dampers   119
       Ductwork   119
How to Establish the Maintenance Program for Air Handlers   120
Five Operating Practices for Air Handling Systems to Enhance Energy
    Conservation   124
       Scheduled Operation   124
       Optimum Start–Stop Timing   125
       Day–Night Setback   125
       Economizer Cycle   126
Summary   126

CHAPTER 9. SERVICE HOT WATER SYSTEMS   127

Service Hot Water System Energy Maintenance Considerations   127
Service Hot Water Requirements   128
Two Service Hot Water Systems   131
       Storage Systems   131
       Tankless Systems   132
Elements of the Service Hot Water System   132
       Water Heaters   133
       Storage Tanks   133
       The Distribution System   134
       End Uses of Service Hot Water   135
Implementing the Energy Maintenance Program   137
Summary   138

SECTION 3. ELECTRICAL SYSTEM OPERATION AND
MAINTENANCE PRACTICES TO IMPROVE ENERGY
EFFICIENCY

CHAPTER 10. LIGHTING SYSTEMS   143

Building Lighting System Maintenance Considerations   143

Lighting Terminology   145
Efficiency of Light Sources   147
    Incandescent Lamps   147
    Mercury Vapor Lamps   149
    Fluorescent Lamps   150
    Metal Halide   151
    High-Pressure Sodium   151
    Low-Pressure Sodium   152
How to Determine Lighting Requirements   152
  Worker Age   153
  Speed and Accuracy   153
  Contrast   153
  Recommended Lighting Levels   153
How to Develop a Lighting Maintenance Program   155
    Step 1: Use Automated Lighting Controls   155
    Step 2: Schedule Lamp Replacement   159
    Step 3: Clean Lamps and Fixtures   161
    Step 4: Inspect System Components   164
Implementing the Maintenance Program   164
Summary   166

CHAPTER 11. ELECTRIC MOTORS   166

Electric Motor Energy Maintenance Considerations   166
  Motor Efficiency Considerations   167
  Higher-Efficiency Motor Standards   168
  Building Motor Maintenance Programs Today   170
How to Develop a Motor Maintenance Program   171
  Step 1: Develop a Motor Inventory   172
  Step 2: Record Motor Maintenance History   172
  Step 3: Determine Motor Size   174
Elements of the Motor Maintenance Program   178
  System Voltage Testing   178
  Routine Inspection   180
  Lubrication   180
  Insulation Testing   180
  Alignment   181
  Vibration Measurement   182
Implementing the Motor Maintenance Program   182
Motor Replacement Program   183
The Impact of Cycling on Motor Life   184
Summary   185

SECTION 4. HOW TO REDUCE THE IMPACT OF AGING ON THE ENERGY EFFICIENCY OF COMPONENTS OF THE BUILDING ENVELOPE

CHAPTER 12. DOORS AND WINDOWS   189

Door and Window Energy Maintenance Considerations   189
      Door Construction   191
      Door Hardware   191
      Window Construction   192
Three Ways Doors and Windows Contribute to the Facility's Energy Use   193
      A Special Warning About Overhead Doors   198
How to Develop a Door and Window Maintenance Program   198
      Step 1: Develop the Door Inventory   199
          Three Door Maintenance Activities   199
      Step 2: Develop the Window Inventory   201
          Three Window Maintenance Activities   201
Implementing a Door and Window Maintenance Program to Reduce Energy
    Requirements   204
      The Repair/Replace Decision   206
Summary   207

CHAPTER 13. ROOFS   208

Facility Roof Energy Maintenance Considerations   208
Forces Acting on the Roof   209
      Heat and Radiation Forces   209
      Chemical Forces   210
      Physical Forces   211
          Foot Traffic   211
          Ponding   211
          Wind   212
          Movement   212
How to Develop a Roof Maintenance Program   213
      Step 1: Develop a Roof Inventory   214
      Step 2: Inspect Roof   214
      Step 3: Conduct Roof Tests   215
Six Types of Roofs   217
Built-up Roofs and Their Maintenance Requirements   217
      Common Built-up Roof Problems   218
      Built-up Roof Inventory   220
      Built-up Roof Maintenance Activities   220
          Inspecting Built-up Roofs   220

Testing of Built-up Roofs   225
Repairing of Built-up Roofs   225
Single-Ply Roofs and Their Maintenance Requirements   225
    Common Single-Ply Roof Problems   227
    Single-Ply Roof Inventory   228
    Single-Ply Roof Maintenance Activities   231
        Inspecting of Single-Ply Roofs   231
        Testing of Single-Ply Roofs   231
Modified Bitumen Roofs and Their Maintenance Requirements   231
    Common Modified Bitumen Roof Problems   235
    Modified Bitumen Roof Inventory   236
    Modified Bitumen Roof Maintenance Activities   236
        Inspecting of Modified Bitumen Roofs   236
        Testing of Modified Bitumen Roofs   236
Metal Roofs and Their Maintenance Requirements   242
    Common Metal Roof Problems   242
    Metal Roof Inventory   243
    Metal Roof Maintenance Activities   243
Shingle and Tile Roofs and Their Maintenance Activities   248
    Common Shingle and Tile Roof Problems   248
    Shingle and Tile Roof Inventory   249
    Shingle and Tile Roof Maintenance Activities   249
Implementing a Roof Maintenance Program   249
Guidelines for Determining Whether to Repair, Replace, or Recover a Roof   252
Summary   253

CHAPTER 14. FOUNDATION AND EXTERIOR WALLS   254

Foundation and Exterior Wall Energy Maintenance Requirements   254
    Thermal Properties of Foundation and Exterior Walls   254
    The Impact of Neglected Maintenance   255
Three Types of Exterior Wall Construction   257
    Wood Frame Construction   257
    Masonry Construction   257
    Metal Construction   258
Two Types of Foundation Wall Construction   258
Forces Acting on Exterior Walls and Foundations   259
    Heat and Radiation Forces   259
    Water Penetration   260
    Physical Forces   260
How to Develop the Foundation and Exterior Wall Maintenance Program   261
    Step 1: Develop the Foundation and Exterior Wall Inventory   261
    Step 2: Inspect Foundation and Exterior Walls   261

Step 3: Test for Damage   264
Implementing a Foundation and Exterior Wall
   Maintenance Program   264
Summary   269

**SECTION 5. HOW TO ESTABLISH A COMPREHENSIVE AND
COST-EFFECTIVE ENERGY MAINTENANCE PROGRAM**

CHAPTER 15. HOW TO SELL THE ENERGY MAINTENANCE
PROGRAM TO SENIOR MANAGEMENT   273

A Recommended Approach for Obtaining Approval of the Energy
   Maintenance Program   274
     Speaking the Language   275
     Perform Energy Comparisons   276
     Implement a Pilot Program   277
     A Case Example   278
Information Senior Management Needs Prior to Commitment   279
How to Ensure Ongoing Commitment for Your Program   281
     Presenting Status Reports   281
Summary   282

CHAPTER 16. GUIDELINES FOR PLANNING AN ENERGY
MAINTENANCE PROGRAM IN YOUR FACILITY   283

Planning the Implementation   284
Setting Maintenance Priorities   286
How to Establish Program Goals   287
     Step 1: Conduct a Complete Inventory   289
     Step 2: Estimate Man-Hour Requirements   289
     Step 3: Review Equipment Purchase Orders   289
     Step 4: Select Systems and Components   290
Summary   292

CHAPTER 17. HOW TO IMPLEMENT AN ENERGY MAINTENANCE
PROGRAM: THREE APPROACHES   293

How to Effectively Use In-House Programs   294
     The Costs and Benefits of Using In-House Personnel   294
     When to Use In-House Personnel   295
     How to Implement Energy Maintenance Using In-House Personnel   296
How to Effectively Use Contracted Programs   297
     The Costs and Benefits of Using a Contracted Program   300
     When to Use Contracts   301

How to Implement Energy Maintenance Using Contracts   302
How to Select the Right Contractor   303
How to Effectively Use Hybrid Programs   304
The Benefits of Using a Hybrid Program   304
How to Implement Energy Maintenance Using Conracts and
In-House Personnel   306
How to Select an Implementation Option   307
Summary   307

CHAPTER 18. HOW TO MEASURE ENERGY MAINTENANCE
PROGRAM EFFECTIVENESS   308

Energy Accounting   309
Energy Use per Square Foot   311
Energy Use per Unit of Production   312
Energy Use per Heating Degree Day   314
Statistical Models of Energy Use   314
Maintenance Performance Tracking   318
Breakdown Frequency   319
Maintenance Requests   320
Breakdown Maintenance Overtime Hours   322
Availability Factor   324
Benchmarking Energy and Maintenance Performance Data   326
Approaches to Benchmarking   328
Steps in Benchmarking   329
Summary   330

APPENDIX   331
Glossary of Energy Terms   331
Energy Conversion Factors   332

Index   339

List of Figures

4.1	Average Annual Energy Use by Building Type	31
4.2	Average Annual Energy Use by Operating Hours	32
4.3	Average Annual Energy Use by Climate	34
4.4	Average Annual Energy Use by Building Age	35
4.5	Average Annual Energy Use by Exterior Wall Material	36
4.6	Average Annual Energy Use by Percent Window Glass	38
6.1	Chiller Log	56
6.2	The Impact of Fouling Factor on Chiller Efficiency	58
6.3	Daily and Weekly Maintenance Activities for Vapor Compression Chillers	63
6.4	Annual Maintenance Activities for Vapor Compression Chillers	64
6.5	Daily and Weekly Maintenance Activities for Absorption Chillers	67
6.6	Annual Maintenance Activities for Absorption Chillers	68
6.7	Weekly and Monthly Maintenance Activities for Cooling Towers	77
6.8	Annual Maintenance Activities for Cooling Towers	78
6.9	Weekly and Monthly Maintenance Activities for Centrifugal Pumps	80
6.10	Annual Maintenance Activities for Centrifugal Pumps	81
7.1	Cost Savings Through Efficiency Improvement	85
7.2	Boiler Log	87
7.3	Boiler Blowdown Requirements	92
7.4	Boiler Flue Gas Oxygen Content Resulting from Excess Combustion Air	94
7.5	Boiler Flue Gas Carbon Dioxide Content Resulting from Excess Combustion Air	95
7.6	Daily and Weekly Boiler Maintenance Activities	97
7.7	Monthly Boiler Maintenance Activities	98
7.8	Annual Boiler Maintenance Activities	99
7.9	Condensate Monitoring	104
8.1	Recommended Ventilation Rates	114
8.2	Air Handler Inventory Data	121

8.3	Monthly Air Handler Maintenance Activities	122
8.4	Annual Air Handler Maintenance Activities	123
9.1	Average Service Hot Water Requirements for Various Building Types	129
9.2	Recommended Service Hot Water Temperatures	130
9.3	Service Hot Water Requirements	136
9.4	Monthly Service Hot Water System Maintenance Activities	139
9.5	Annual Service Hot Water System Maintenance Activities	140
10.1	Efficacies for Common Lamps	146
10.2	Color Characteristics for Common Lamp Types	148
10.3	Recommended Illumination Levels	154
10.4	Lighting Survey	156
10.5	Survival Rate for Fluorescent Lamps	162
11.1	Performance of Standard and High-Efficiency Motors	169
11.2	Motor Inventory Data	172
11.3	Motor Test Data	175
11.4	The Effect of Part-Load Operation on Motor Efficiency and Power Factor	176
11.5	Motor Maintenance Activities	179
12.1	U-Values for Wood and Steel Doors Without Glazing	195
12.2	U-Values for Door and Window Glazing	196
12.3	Typical Door and Window Glazing Coefficients	197
12.4	Door Inventory	200
12.5	Door Maintenance Activities	202
12.6	Window Inventory	203
12.7	Window Maintenance Activities	205
13.1	Built-up Roof Inventory	221
13.2	Built-up Roof Condition	223
13.3	Single-Ply Roof Inventory	229
13.4	Single-Ply Roof Condition	232
13.5	Modified Bitumen Roof Inventory	235
13.6	Modified Bitumen Roof Condition	239
13.7	Metal Roof Inventory	244
13.8	Metal Roof Condition	246
13.9	Shingle and Tile Roof Inventory	250
13.10	Shingle and Tile Roof Condition	251
14.1	Foundation and Exterior Wall Thermal Conductivity	256

14.2	Foundation Wall Inventory	262
14.3	Exterior Wall Inventory	263
14.4	Foundation Wall Condition	265
14.5	Exterior Wood Frame Wall Condition	266
14.6	Exterior Masonry Wall Condition	267
14.7	Exterior Metal Wall Condition	268
17.1	Energy Maintenance Job Estimate	298
17.2	Maintenance Job Scheduling	299
17.3	Contractor Qualifications	305
18.1	Facility Energy Use Log	310
18.2	Facility Btu per Square Foot Worksheet	313
18.3	Facility Btu per Unit of Production Worksheet	315
18.4	Facility Btu per Heating Degree Day Worksheet	316
18.5	Maintenance Breakdown Log	321
18.6	Maintenance Requests from Occupants	323
18.7	Breakdown Maintenance Overtime Log	325
18.8	Availability Factor Worksheet	327
A.1	Energy Conversion Factors	332

Section 1

Energy Maintenance in Facilities

Chapter 1

The Energy–Maintenance Connection

When facility managers are asked what they are doing to promote energy management in their facilities, most will speak of investments in extensive and expensive construction and renovation projects. Replacing windows, building cogeneration plants, installing high-efficiency chillers, retrofitting fluorescent fixtures with electronic ballasts, constructing thermal storage systems; all are frequently cited projects being implemented to reduce energy use within facilities. Rarely does the facility manager talk about maintenance activities as a means of reducing energy use. Energy projects, construction, and renovation is where the excitement and attention are. Maintenance is boring.

Unfortunately, many of these facility managers misunderstand the significance of sound maintenance in an energy management program. Successful energy management requires the implementation of a thorough maintenance program. Properly maintaining building components and systems reduces energy requirements for those components and systems. The two are so intertwined that it is impossible to separate activities that promote energy management from those that promote sound maintenance. It is no coincidence that the most successful energy management programs of the 1970s, 1980s, and 1990s were found in the best managed and maintained facilities. Those facility managers recognized that energy conservation, while important to the economic health of the facility, was only another tool to help manage the operation of the facility. Ignoring maintenance in favor of high-cost energy conservation projects will reduce the impact of other energy management efforts.

For example, a large facility distributed steam from a central plant to approximately 100 buildings through an underground system. While portions of the system had been upgraded as the facility grew over the years, much of it was more than forty years old and in poor condition. Less than 20 percent of the condensate generated was returned to the central plant due to corroded and plugged piping. Pipe insulation was severely deteriorated, or in some cases, nonexistent. Pipe flanges and valve packings leaked.

3

Manholes regularly filled with water, creating large plumes of vapor and allowing water to further deteriorate piping insulation. More than three-fourths of the steam traps serving the distribution system were not working properly; many had failed to open. Tests conducted on the system showed that 30 to 40 percent of the annual steam production from the central plant was lost due to leaks, bad insulation, failed traps, and lack of a condensate return system.

To reduce energy costs, the facility decided to invest heavily in a cogeneration system. The savings produced by the system were expected to recover the capital investment over a period of seven to ten years. While the higher efficiency of the new central, cogeneration equipment reduced energy costs, it did nothing to address the inefficient distribution system. Thirty to 40 percent of the annual steam produced by the new cogeneration system was still lost due to leaks, bad insulation, failed steam traps, and dumped condensate. Adding new, more efficient sources of steam to an inefficient distribution system may save money but it is not energy management.

In contrast, if the facility had implemented a comprehensive steam trap maintenance program, repaired known leaks, taken steps to prevent water damage, and repaired the damaged portions of the condensate return system, they could have achieved comparable savings for only a fraction of the initial investment.

While there is no doubt that energy conservation projects, if properly conceived, planned, and implemented, will help to reduce energy use, so will good maintenance. But unlike energy conservation projects, good maintenance can reduce energy use without requiring a major investment of capital. Surprisingly, the energy conservation payback for maintenance activities is typically measured in months, not in years as is the case for most energy conservation projects.

Why then are maintenance activities so often overlooked or ignored in energy management when the rewards are so great? There are at least three major factors that contribute to the misunderstanding of the relationship between energy management and sound maintenance; the attitude of the organization toward maintenance, the history of energy management, and the nature of maintenance activities.

A TRADITIONAL VIEW TOWARD MAINTENANCE

In spite of the efforts of maintenance managers to educate organizations, maintenance continues to be viewed as a necessary evil, a drain on the resources of the organization that could be better used elsewhere. Even the use of the term maintenance helps to reinforce this belief. Maintenance means keeping things in the existing state. One would be hard-pressed to find other areas within an organization where the goal was to maintain the status quo. Nearly all other areas are funded to change things; investments are made, rates of return are established, acquisitions are made, personnel are hired, the organization is moved forward.

There are no comparable developments apparent in the field of facility maintenance. Great sums of money, typically 5 to 15 percent of an organization's budget, are poured into maintenance every year with little visible results. At best, things are kept as they were. The status quo is maintained. But maintaining the status quo does not generate excitement or support. If a maintenance department is doing its job properly, the work will be transparent to the organization. Perhaps that helps explain why maintenance historically has been the first area to be cut during rough financial times for the organization.

Contributing to the negative attitude toward maintenance is the common belief outside maintenance departments that if something isn't broken, don't fix it. That belief leads to the conclusion that if something does not need fixing, the money can be better spent elsewhere. In spite of the understanding by maintenance departments of the value of preventive maintenance, few organizations today have a truly comprehensive preventive maintenance program in place. Unfortunately, what many maintenance managers promote as preventive maintenance is really little more than routine maintenance. For example, in some organizations, the replacement of light bulbs after they have burned out is performed by preventive maintenance crews and recorded as a preventive maintenance activity.

This attitude of not fixing things that aren't broken is reinforced every time maintenance is deferred. Chances are that unless the system has become severely deteriorated, maintenance can be deferred and the system will keep on running, at least for the short term. The organization sees this, so it simply concludes that maintenance they deferred was not necessary after all, reinforcing the attitude that deferring maintenance is a sound practice. Only when things actually stop working are they willing to accept the need for maintenance. What they do not see is the deterioration that is taking place as the result of aging and deferring maintenance and the impact it is having on system efficiency—deterioration that is only accelerated by a lack of maintenance. And systems that are not operating at their peak efficiency are costing the organization through wasted energy.

When building systems are new, assuming that they have been properly installed and set up, their energy efficiency is at its peak. Heat exchanger surfaces are clean and free of scale. Outside air dampers seal tightly and operate smoothly. Temperature control systems are calibrated and operate within the established limits. Insulation is dry and tightly adhered. Lighting system lamps are new and the diffusers are clear and clean.

With time and normal use, efficiency goes downhill. Scale and dirt accumulate on heat exchanger surfaces. Air damper seals leak and actuators stick. Controllers drift and go out of calibration. Insulation is damaged, becomes wet, or separates from the surfaces it is designed to protect. The light output from fixtures decreases and diffusers become cloudy and dirty. The result is decreased energy efficiency.

This process of slow deterioration in building systems is a natural process resulting from simple aging and use. The rate of deterioration depends on the quality of the systems and components installed, how those systems are operated, and how well those systems are maintained. Although the facility manager has little control over the quality of the installed systems, he or she does have some control over how those systems are operated. Facility managers exercise even greater control over how well those systems are maintained. With good maintenance, the rate of deterioration, particularly in those areas that impact energy performance, will be reduced or halted.

Unless facility managers are constantly reminded of the negative impact of deferred and neglected maintenance, the organization will continue to downgrade or ignore the maintenance aspects of energy management.

THE FOCUS ON ENERGY PROJECTS

The idea that conserving energy was going to cost big money developed almost immediately with the oil embargo of 1973. When prices rapidly rose, organizations immediately implemented a two-phase approach to energy conservation. The first phase consisted of implementation of quick fixes that would start saving energy immediately without requiring a significant investment of capital—turning down thermostats, reducing lighting levels, reducing ventilation rates, and so forth. While the results of these quick fixes produced savings, it was widely believed that a second phase was needed to achieve even greater savings. This second phase consisted of projects that, unlike the quick fixes, required a significant investment of capital to either replace or overhaul the existing inefficient systems installed in buildings. As a result, projects were justified on the basis of simple payback or return on investment.

In the years following 1973, more and more emphasis has been placed on projects as the means of conserving energy. Single glazed windows have been replaced with dual or triple glazed units, some with high-tech coatings to further reduce energy losses. Constant volume, dual duct, and reheat systems have been replaced with variable volume designs. Constant volume heating and cooling pumping stations have been equipped with variable frequency drives and control systems that match system flow rates with the demand being placed on the system. Ice storage systems have been constructed to make use of lower, off-peak utility rates. In all, billions of dollars were spent on projects to reduce energy use.

The emphasis on energy projects as the solution to high energy costs continues today. With the continuing belief that all of the easy, low-cost measures have already been implemented, facility managers continue to push large and costly projects as the only means of advancing energy conservation. Projects continue to grow in size, complexity, and cost. Many have grown so costly that they cannot be implemented without outside assistance in the form of utility rebates or shared energy cost savings.

This belief that all of the quick fixes have already been implemented is widely accepted today as fact. Further, it is believed that if organizations are to pursue additional cuts in energy use, even larger and more expensive projects must be implemented. Trade journals, company news letters, and other publications regularly report on energy conservation projects that are being completed by organizations. Those projects typically cost hundreds of thousands of dollars to implement and involve the use of high technologies or new energy conservation strategies. Mundane maintenance activities that reduce energy use do not get comparable coverage even though the net result is the same. This simply serves to reinforce the belief that energy conservation requires the implementation of big ticket projects.

LACK OF GLAMOUR

For nearly ten years now, energy conservation has been largely ignored by the media, public officials, corporate officers, and the general public. This absence of attention is a result of a lack of long-term energy concerns. Fuel supplies have been ample and prices have remained relatively low. Only during the temporary rise in gas prices during 1996 was any attention paid to energy as an issue, and even then it was treated as a short-term anomaly resulting from a combination of factors that soon went away. Energy conservation has fallen out of favor and has become secondary to other issues.

This lack of focus on energy has changed how organizations approach the issue of energy conservation. During the 1970s and early 1980s, conserving energy was practically a requirement for survival. High costs and limited availability dictated that facility managers do all that they could to reduce energy use. Even though the economy was not good, funds were made available for energy conservation activities as it was understood that the money being spent was an investment—an investment that not only helped ensure the survival of the organization, but also provided a positive rate of return. Economics and survival were the driving forces behind those programs.

Today the situation has changed. While economics is still an important issue, the survival aspects of energy conservation are not. Nobody today believes that they face curtailments or a total shutdown of their operations as the result of an energy supply problem. Energy is abundant and the future looks good because there is no bad news.

Even the economic factors no longer carry the same weight that they once did. Competition for funding has increased with other activities—activities that promise an even higher rate of return. Routine energy conservation activities simply do not capture the attention of corporate officers. Not all of the blame lies at the corporate level; at least a portion of it must be shared by the facility manager. If maintenance activities for energy conservation are to compete for funding, they need an advocate to promote their benefits to the organization. Simply identifying the activities and requesting the funding is insufficient. Maintenance activities for energy conservation must be promoted in exactly the same manner as other programs seeking funding, including identification of costs, benefits, and return on investment. Few maintenance managers have done this.

Without the advocate providing the necessary information to promote energy conservation through maintenance, maintenance does not stand a chance. Therefore, maintenance managers have turned to other means of promoting energy conservation: projects. In the eyes of the corporation, construction projects are more glamorous than maintenance. Maintenance managers soon realized that if they were to get any funding for energy conservation, they would have to pursue the types of activities that the organization was interested in and was willing to support: energy conservation projects. The reality of the situation today is that a facility manager is far more likely to receive approval for a multi-million-dollar ice storage system than for a $10,000 per year chiller maintenance program, even though both will reduce energy costs.

Part of the reluctance to fund maintenance activities is the result of a lack of understanding of how maintenance activities promote energy conservation. Few outside of the maintenance department understand how a systematic maintenance program on a chilled water plant can save energy by such routine activities as purging air from refrigerant systems or cleaning scale from heat exchangers. It is much easier to understand how the ice storage system will reduce costs by shifting the load to off-peak hours.

Lack of understanding is only part of the problem. Realistically, maintenance activities simply do not generate good press. Large, expensive energy projects do. In this age when organizations are very image conscious, activities are carefully crafted to promote a positive image of the organization. The ice storage system is something that can be seen and shown off to the press and the public. A clean chilled water system cannot.

Large expensive projects for energy conservation are selected in part because they can be used to help promote a positive image of the organization. Through them, an organization can help to promote the image of being concerned about issues of energy use and the environment. Projects help to promote that image, maintenance does not. Projects, especially high-tech ones, get headlines. Maintenance does not.

THE ENERGY MANAGEMENT ROLE OF MAINTENANCE

A comprehensive energy management program will use both energy projects and maintenance activities to help minimize energy use. Neither can do it alone. Good maintenance will enable the facility manager to operate the building systems and components at their peak energy efficiency. That efficiency, however, will be limited by the operating characteristics of those systems and components. Further increases in efficiency require implementation of a project to replace the system or component with a newer-generation, more efficient one.

For example, a ten-year-old centrifugal chiller that has received only routine maintenance will typically have an actual full-load operating efficiency of approximately 1.0 kW per ton. A comprehensive maintenance program, consisting of an annual or semi-annual tube inspection and cleaning, balanced water treatment, and refrigerant leak monitoring, would typically result in an increase in full-load operating efficiency to 0.7 to 0.8 kW per ton, producing an improvement in seasonal operating efficiency and energy savings of 20 to 25 percent. For a medium to large chiller plant, the cost of the maintenance would be recovered through energy savings in just a few weeks. However, new chillers now are available that have a full-load operating efficiency of approximately 0.55 kW per ton, producing even greater savings.

Which option is better for a given facility? Both require an investment. Both save energy. The chiller replacement energy project option saves more energy but also costs more to implement. The maintenance option provides a quicker payback. Which option is better for a facility? It depends on the particulars of the installation. If the chiller is fairly new and in good operating condition, establishing the maintenance program would be the most cost-effective option. Similarly, if budgets are tight and immediate results are needed, maintenance will provide an immediate and cost-effective return. If, however, the chiller is in poor operating condition and is approaching the end of its expected normal useful life, replacement with a high-efficiency unit will be the best option.

Even then, performing at least some of the maintenance tasks will provide a positive return, particularly if the new chiller will not be installed for another six months or longer.

Even when energy conservation projects are being promoted by the organization, maintenance is important. Today's energy conservation projects are frequently designed around newer, high-tech systems and components. These systems and components have tighter operating tolerances and place increased demands on maintenance personnel. Both routine and preventive maintenance activities are increased just to keep the systems operating as they were designed to. Without these maintenance activities, the systems would rapidly lose their effectiveness as tools to promote energy efficiency. Without maintenance, today's new, energy efficient systems become tomorrow's energy hogs.

For example, in a facility that required year-round air conditioning, an organization installed a plate-and-frame heat exchanger in parallel with the chiller. Chilled water was piped through one side of the heat exchanger while cooling tower water was piped through the other. According to the design, the heat exchanger would provide free cooling to the facility whenever the outside air temperature was below 45 degrees. The energy savings were sufficient that the system would pay for itself in less than two years.

Although the system performed well, it placed additional requirements on the maintenance department in both materials and labor. About three years after the system had been installed, part of the control system failed. Even though the cost of replacing the controls was less than the savings produced by the system during one year's operation, the organization was not willing to invest the money in maintenance activities, and the system was abandoned.

Even though maintenance is essential to energy management, maintenance by itself is not sufficient. Energy management requires a balance between maintenance and projects. Just as it made no sense to install the heat exchanger, then let it fail due to a lack of maintenance, it is equally as illogical to keep pouring maintenance funds into systems that are inefficient by nature and cannot be improved through maintenance. Group cleaning and group relamping may be good maintenance practices but they are not good energy management. Good energy management would dictate that incandescent lighting systems be replaced with higher-efficiency light sources. In this case, upgrading to a higher-efficiency source would also be good maintenance as the maintenance requirements are less for high-efficiency sources than they are for incandescent lamps.

PROMOTING MAINTENANCE AS GOOD ENERGY POLICY —————————

Making maintenance an integral part of the energy management program for a facility is the responsibility of the maintenance manager. As maintenance manager, you are the one closest to the energy-using equipment and systems and have the best understanding of the impact that a lack of maintenance has on those systems and their energy use. But you cannot simply say that improving maintenance through increased spending will save energy. The competition for the organization's resources is too high to request funding on a simple act of faith.

If you are to compete with other programs for funding, your case must be built around costs and benefits. The potential for energy conservation savings must be identified and quantified. Savings must be tied to specific maintenance activities. The costs of performing those maintenance activities must be quantified, particularly when they go beyond what the organization has come to accept as the norm.

In some cases you may not be able to quantify the energy savings associated with a particular maintenance activity. In those instances, it will be beneficial to emphasize what would happen if maintenance is not performed, in terms of both energy efficiency and reliability. For example, it is very difficult to quantify the energy savings that would result from an annual testing and calibration of the operators and controllers in a building's temperature control system. Simply assuming that a certain percentage of the units are not operating properly is insufficient as there is no means of backing up the assumptions, unless there are data from previous calibration efforts. Instead, promote the effort by citing the importance of calibration of the controls to proper operation of the system as well as to energy efficiency. Discuss the long-term cost of allowing the controllers to operate out of calibration. If previous calibration efforts have been completed, cite the conditions found during those efforts.

In addition to providing the energy cost benefits of maintenance, you must help focus the attention of the organization on maintenance and what it means to operations. Remember, maintenance is generally overlooked by the organization. Even those in the organization who are aware of the maintenance operation seldom realize that maintenance is more than just fixing things that are broken. Focus attention on how maintenance helps support the overall mission of the organization. Show how good maintenance improves operating efficiency and makes system operation more reliable. Promote maintenance as an investment.

Help to fight further reduction in maintenance budgets by identifying the cost of deferred maintenance. Be specific. When maintenance is deferred, it costs the organization in reduced reliability, efficiency, and equipment operating life. Stress those costs and what they mean to the organization in both the short and long term. Show how deferring maintenance as a means of cost cutting is short sighted.

Finally, promote the accomplishments of maintenance when it comes to operating efficiency. Too often maintenance managers prefer to work quietly behind the scenes, forgetting that they are competing with other areas within the organization for funding. Cite tasks that have been completed and how they impact the operating efficiency of the energy using systems. For example, if a chiller has been recently cleaned or overhauled, discuss what was done, why it was done, what was found, the cost savings are going to be produced by the effort, and what would have happened if the maintenance had been deferred.

If you have performed maintenance activities that have improved the operating efficiency of a component, system, or even an entire facility, share the results with the organization. Compare the efficiencies that you have achieved with published norms or benchmarks used by others to rate energy efficiency. Show how maintenance has helped the organization to become more efficient and save operating funds.

Chapter 2

How Energy Management Is Traditionally Performed by Organizations: A Five-Phase Process

Energy management is a process. It is a well structured process that is both technical and managerial in nature. Using techniques and principles from both fields, energy management monitors, records, investigates, analyzes, changes, and controls energy use within organizations. Its objective is to see that all energy using systems within the organization are supplied with all of the energy that they need, when they need it, in the form that they need, at the lowest possible cost, and that the energy supplied to those systems is used as efficiently as possible.

Although the phrase "energy management" is often used interchangeably with "energy conservation," the two are not the same. Energy management is much broader than energy conservation in both what it strives to accomplish and the techniques it uses to accomplish its goal. While both seek to reduce overall energy costs to the organization, energy conservation focuses on individual measures or activities that can be implemented to reduce energy use. In contrast, energy management looks at the entire process of energy use. Energy conservation programs tend to consist of a number of one-time efforts, while energy management is an ongoing activity.

There are three major elements to an energy management program: planning, implementation, and control feedback. Planning begins with the understanding that energy is a resource to be managed just like all other resources available to the organization. It establishes realistic goals for energy use levels within the organization, identifies programs needed to achieve those goals, and develops procedures to monitor performance and progress toward achieving those goals.

Implementation requires a commitment from the highest levels in the organization. Energy management impacts all operations of the organization and competes with many other operations for limited resources. Without this commitment of backing, personnel, and financing, the program will quickly be rendered ineffective.

11

Implementation continues with the development of an energy database that identifies how much of what type of energy is being used when and where in the facility. The database is used to help identify energy conservation measures that will assist the facility in meeting the goals established during the planning phase. Information in the database takes two forms: metered data and the results of an energy audit. Metered data is particularly useful in large, multi-building complexes where data is recorded on a building-by-building basis. As the database grows with time, the information can be used to track performance of the energy using systems within one building as well as to provide a basis for comparison between similar buildings.

Metered data is also useful at the system level. By installing meters on a single system, the performance of that system can be tracked over time. Modifications to that system can be evaluated to determine which energy conservation activities work and which do not.

The energy audit is an effective tool for building a foundation of energy conservation opportunities. The energy audit identifies and evaluates all energy using systems and components in a facility, and attempts to quantify the portion of facility energy use that they impact. By completing the energy audit, facility managers will identify both the largest energy users in the facility and the largest energy saving opportunities.

The control feedback is needed to evaluate how well the energy management program is achieving its goals. Energy conservation measures are implemented based on certain assumptions about how the facility is operated and what impact the measures will have on energy use. Without good feedback, there is no way to confirm that the assumptions were correct or that the new or modified system is performing as intended.

Feedback builds on the energy database established by the energy management program. It updates energy use figures and provides a very accurate determination of the program's effectiveness.

Feedback also provides the data necessary to keep the energy management program going. Too often management initiatives die a slow death due to the lack of ongoing support. Programs start out strong, but quickly fade as management focus and support move on to new areas. Feedback from the energy managers on the status of the program, how well it is achieving its goals, and hard data on savings that have been achieved are a very positive means of keeping attention focused on energy management.

THE SUDDEN NEED FOR ENERGY MANAGEMENT

Although a few energy-intensive industries had been using the techniques of energy management prior to 1973, energy management was for the most part unheard of until then. Energy costs were low, particularly relative to other costs of doing business. There was a widespread belief that our energy resources were virtually limitless. Energy shortages were unthinkable. Conserving energy was equated with reducing the standard of living for the general public, and cutting the productivity and output for organizations. The whole concept of doing more with less energy was so foreign to business that nobody understood the potential benefits that could be derived from energy management.

Everything changed when Arab members of the Organization of Petroleum Exporting Countries (OPEC) banned the export of oil to the United States and the Netherlands on October 19, 1973. The embargo was in retaliation for support and military aid given to Israel during the October Middle East War. Additionally, OPEC announced a 25 percent cut in the production of crude oil. The combination of the embargo and the announced reduced production levels created panic and set off a round of wild bidding by nations that were net importers of oil. Just prior to the embargo, crude oil was selling for $3.29 a barrel. In early 1974, its price had reached $11.06. The perception of governments, companies, and the general public was that there was an actual physical shortage of energy even though less than 7 percent of the world's oil supply was impacted by the actions of OPEC.

The impact of the embargo was felt almost immediately in the United States. Gasoline stations were ordered closed on Sundays. A standby program for gasoline rationing was established. Outdoor decorative lighting was ordered turned off during the holiday season. An energy czar was appointed. The national speed limit was lowered to 55 miles per hour. Trips were measured in tankfulls. Car values were expressed in miles per gallon.

Perhaps the most serious impact of the embargo was the triggering of a worldwide recession. World economies were hit with double-digit inflation, while their economic growth stagnated and unemployment soared. Few other events in world history have had so broad and so long lasting an impact.

One positive aspect of the embargo was that it served as a wake-up call to the world. People recognized that oil was a limited resource. Given the rate at which the world was using oil, and the rate at which new oil fields were being developed, it was realized that the world would run out of oil. Even without considering the impact that political factors and instability have on oil supplies, it was agreed that there was a real need to conserve energy. Add in those political factors, and the situation became even more serious. Energy shortages were no longer considered to be some doomsayer's wild prophecy; they were real, they were already here, and unless we did something about energy use, they would impact our standard of living.

In the years following the oil embargo, energy management has gone through four distinct phases, and is now entering a fifth. Understanding where we are and how we got here is important if we are to build on successes and avoid repeating past mistakes.

Phase I: Quick Fixes

Faced with rapidly escalating costs and the prospect of closings resulting from energy shortages, facility managers responded by implementing a round of energy conservation measures. These first measures, known as quick fixes, required little planning or capital to implement. Thermostats were lowered during the heating season and raised during the air conditioning season. Outside air dampers in building heating, ventilating, and air conditioning (HVAC) systems were closed off. Fluorescent tubes were removed from fixtures. Lighting systems were placed on timers. Time clocks were added to building HVAC equipment. Building shells were sealed up.

Most quick fixes had a positive impact on energy use, although some did not. Those that did not work properly generally failed because of a lack of understanding and unintended consequences. For example, in some building HVAC systems, resetting the building's thermostat to a lower setting during the heating season did nothing more than cause the system to use more outside air to lower the temperature. In other system designs, it actually increased energy use by causing the building chillers to come on line.

With time and experience, facility managers did resolve the operational problems, and quick fixes did produce significant energy and cost savings. More importantly, they demonstrated that it was possible to reduce building energy use while maintaining a suitable environment. Quick fixes laid the necessary groundwork for developing additional energy conservation measures, measures that would require time and money in order to implement.

Phase II: Energy Projects

Quick fixes were generally justified simply because common sense said that they saved energy. Thermostats were turned down during the heating season because lower temperatures meant less energy would be used to heat the building. Nobody was certain how much energy would be saved by the lower setting, but common sense said that it would save energy.

Once a fairly wide range of quick fixes had been implemented, facility managers came to realize that if additional savings were to be achieved, it would require the implementation of energy conservation activities that required both a significant amount of time to put in place and a significant investment of capital. As a result, the emphasis shifted from quick fixes to energy projects.

The same basic common-sense plan was followed with many of the early energy conservation projects. Common sense said that projects such as adding insulation to a building or replacing windows saved energy. Therefore nearly any project that made sense was eligible for funding. As long as money was available, they were funded. Saving energy had become the primary goal; an end unto itself.

Unfortunately, in many cases when the energy bills were reviewed after the project had been completed, it became difficult to identify the energy savings produced. It took some time, but facility managers realized that they were simply throwing money at the problem without an understanding of what they were getting in return. Common sense may have worked when the price of the quick fix was low, but energy conservation projects were not low cost. Projects had to be justified in terms of their return on the investment.

Although many different methods were used to calculate the return on investment for an energy conservation project, the most common method was the simple payback. Simple payback was easy to understand and simple to use. It provided a convenient gauge by which the relative energy conservation merits of a project could be measured and compared.

An energy project's simple payback, expressed in months or years, is determined by dividing the project's estimated implementation cost by the estimate of annual savings that it will produce. Projects that had a simple payback period of one year or less were funded from operating funds, often directly from the fuel and utility savings. Projects with longer, simple paybacks required capital investment and were funded through various means depending on the policies of the organization.

When energy management programs were first implemented, in order to be a candidate for funding, an energy conservation projects had to have a simple payback of less than five years. As programs have matured, and as more projects have been implemented, many organizations have set their cutoff point for funding in the range of five to seven years.

Since the oil embargo, a wide range of projects have been implemented in the name of energy management. These projects have included building component replacement, system upgrades, and entire system replacements. Although projects have ranged in cost from a few thousand to several million dollars, the trend over time has been toward increasing project costs.

Phase III: Comprehensive Energy Management

Quick fixes and energy projects provided much needed relief to facility managers. The energy and cost savings they produced helped ensure the continued operation of the facilities through periods of energy shortages and dramatic price increases. However, in spite of these savings, energy prices continued to rise, placing increasing pressures on organizations to do even more to reduce energy costs. To combat these rising costs, organizations developed more comprehensive approaches to energy management—approaches that went beyond programs to reduce energy use to programs that managed energy use. These were the first attempts to actually manage the use of energy.

One of the first steps that organizations took in developing comprehensive energy management programs was to appoint an energy manager. The energy manager typically reported to someone fairly high in the organization, thus helping to ensure that he/she had the authority and support to make the decisions required.

Larger organizations went on to develop a support staff for their energy management function. The staff consisted of a diverse group of technical and managerial personnel, with backgrounds in engineering, accounting, finance, business management, and maintenance. Together with the energy manager, they looked beyond the quick fixes and hit-or-miss energy projects of the past to real management of energy. All aspects of energy use in facilities were examined. Options were developed and evaluated. A plan to manage energy use was developed and implemented. Monitoring systems were put in place to provide ongoing feedback on progress.

Energy management programs that followed this approach achieved great successes, reducing energy use on a per-square-foot or a unit-of-goods-produced basis by 35 percent or more.

An important discovery made by organizations with fully developed energy management programs was the value of good maintenance. Traditionally, most organizations operate in the breakdown maintenance mode. Systems and components are operated until something breaks down. Maintenance is called in and the system is repaired or replaced and put back in service. While the breakdown approach to maintenance may seem a common-sense approach to minimizing maintenance costs, it does not take into consideration the energy saving aspects of good maintenance in general, and preventive maintenance in particular.

Energy managers came to realize that energy using systems are prime candidates for preventive maintenance. Maintaining peak operating efficiency of these systems requires periodic cleaning and adjustments. Without this maintenance attention, system performance deteriorates and energy use increases. Comprehensive maintenance programs demonstrated that one-time maintenance costs were rapidly offset through energy savings, increased system performance, and extended equipment life.

Phase IV: Declining Interest

The 1980s reversed the developments of the 1970s. Energy prices first stabilized then declined. Oil shortages were replaced with a glut of oil. Headlines proclaimed, "In Case You Missed It . . . The Energy Crisis Is Over," and, "The Oil Glut Won't Go Away." Some even questioned if OPEC could survive the glut of oil. Energy management went bust.

The decline continued well into the 1990s. Stable crude oil prices and supplies led to a decrease in gasoline prices. For the general public and public officials, whose attitudes toward energy conservation are largely shaped by availability and pricing of energy, this signaled an end to the need to conserve. The small, energy efficient vehicles of the late 1970s were replaced by fuel inefficient sport utility vehicles and luxury vehicles. Motor-vehicle fuel use has increased nearly 50 percent between 1970 and 1992, in spite of a doubling in new car mileage ratings. The 55-mile-per-hour national speed limit has been eliminated.

Energy complacency also shaped the way businesses operated during the 1980s and early 1990s. Low energy prices lessened the pressure to manage energy use. Funds that normally would have been earmarked for energy management were diverted for other purposes. Energy management programs were scaled back or completely eliminated, the money invested in other efforts deemed more essential. Energy managers and energy management teams were done away with. Energy conservation became associated with anti-growth.

New concerns captured the attention of businesses: downsizing, Total Quality Management (TQM), reengineering. Energy management in comparison was unimportant. Even when events triggered temporary dramatic price increases, such as the Gulf War or the harsh winter of 1995, business analysts passed them off as one-time anomalies that could be ignored.

Phase V: Coming Full-Circle

A number of factors have been combining recently to once again focus attention on energy management. Even though current energy supplies remain plentiful, there is increasing concern over how long they can continue at today's production levels and prices. Two factors in particular give cause to these supply concerns: the increasing reliance on imported oil and dwindling world reserves.

A direct impact of the stable oil supplies and declining costs during the 1980s and early 1990s was the collapse of domestic oil production. Low prices made it less profitable to explore for new oil or even produce from existing wells. As a result, domestic production has declined dramatically. In 1973, just prior to the oil embargo, the United States was importing approximately 38 percent of its oil needs. After fifteen years of fluctuations due to supply problems, price variations, and reduced demand, oil imports have returned to an upward climb. In 1994, oil imports for the first time totaled 50 percent of U.S. needs. This upward trend will continue, particularly with declining oil production in Alaska as the field plays out. By the year 2010, it is projected that the United States will be importing more than 65 percent of its oil needs. High levels of imports aggravate the trade deficit and leave the nation vulnerable to future supply and price disruptions.

Declining world supplies also give cause for concern over future oil supplies. The U.S. Geological Survey estimates that the total quantity of recoverable oil in the entire world is approximately 2,330 billion barrels. If this figure is correct, in less than 100 years we have consumed nearly one-third of this amount.

New discoveries of oil have not kept apace with increasing demand. The world is currently using oil at the rate of approximately 20 billion barrels per year. New discoveries are averaging only 15 billion barrels per year. The difference is coming from known reserves. If new discoveries are expected to close the gap, a supergiant oil field the size of Alaska's Prudhoe Bay would have to be discovered every three years. Since 1968, only three supergiant fields have been discovered.

Another factor contributing to the increase in interest in energy management is a tightening economy. When the economy was booming, energy conservation slipped in priority. Staffing was reduced. Energy conservation funds were diverted to other purposes. But the economy is no longer booming. Cutbacks and downsizing have become the accepted norm. Organizations are looking for any means possible to reduce operating costs. Energy management offers them one effective means of accomplishing this reduction.

This renewed interest in energy management is causing problems for organizations. Few of today's facility managers have any hands-on experience in developing and running comprehensive energy management programs. Those who set up the programs of the 1970s and early 1980s are gone either as the result of normal turnover or reassignment to other programs as energy management programs were scaled back during the era of cheap energy. The result is that many organizations lack experienced personnel who can provide the historical perspective to implement effective energy management.

Compounding the problem is the fact that in the twenty plus years that have passed since the oil embargo, many changes have been made in the way in which facilities are operated. Today, buildings and their energy using systems are more sophisticated.

While there are better systems for monitoring and controlling equipment operation, those systems are much more interrelated than they were twenty years ago. Regulations and concerns over issues such as indoor air quality and CFC use limit some energy conservation opportunities while enhancing the feasibility of others. Quick fixes, which were so effective in reducing energy use following the oil embargo, are not suitable for many of today's building systems.

Organizations are once again recognizing the value of and need for energy management. If they are to be successful, they must understand what worked in the past and why, and what did not work and why it failed. Energy managers must not make the same mistake of throwing money at the problem. Funds are too scarce and the cost of energy conservation measures too high. A more comprehensive approach must be followed.

Instead of randomly funding projects, start with setting up a program to manage energy use. Appoint an energy manager with overall responsibility for developing and implementing the program. Make certain that the effort has top backing in terms of commitment, funding, and authority. Develop an understanding of how and where energy is used in the facility. Benchmark your energy use against other comparable facilities to determine how energy efficient your organization is. Establish reasonable goals for the energy management program. Develop and implement a measurement method to provide feedback that will measure performance. Identify and survey your energy using equipment. Make certain that the equipment you already have is operating as it was intended and as efficiently as possible. Identify your energy conservation options and prioritize their implementation into an energy management plan. Implement the plan as funding permits. Finally, review your progress on an ongoing basis to determine the program's effectiveness.

Chapter 3

Key Incentives for Reducing Energy Requirements in Facilities

It is estimated that commercial and residential buildings account for 35 to 40 percent of the total energy use in the United States. Although new buildings are being constructed according to much tighter energy conservation standards, new buildings are only a small portion of the overall building inventory. For example, nearly 80 percent of the existing commercial building inventory was already in place before 1979. Energy requirements of these facilities typically range from 30 to 40 percent higher than for newer construction. With a mean lifetime of 60 to 70 years for a nonresidential building, there is little hope of retiring and replacing these facilities in the near future. Therefore any reductions in building energy requirements brought about through implementation of energy management will have a significant impact on total energy use.

When OPEC first enacted the oil embargo, facility managers implemented energy management programs to cope with the shortages in energy supplies. As the impact of the embargo grew and energy prices rose sharply, facility managers used energy management techniques to help control their costs. Today, there are three driving forces behind the need for facility managers to implement energy management programs: economics, supply concerns, and environmental issues.

ECONOMIC INCENTIVES FOR ENERGY MANAGEMENT

Economic considerations have always been the primary driving force behind energy management. With the exception of the brief period of time during the OPEC oil embargo when supply temporarily was the key issue, energy management funding decisions have been made almost solely on the basis of economics.

Today, economics is even more important. Increased competition, both nationally and internationally, has forced organizations to undertake major programs to reduce costs. Workforces have been downsized, business processes have been reengineered, and functions have been outsourced—all in an effort to come up with a leaner and more cost-effective operation.

A prime target for some of the cost cutting has been the facility operating budget. Facility budgets are typically viewed as being pure overhead. Any reductions in the operating expenses of a facility translates directly into profit for the organization. Energy management has repeatedly demonstrated itself over the past twenty years to be one of the most cost-effective means for profit improvement—cost reduction. Even modest energy management programs of the 1970s and 1980s were able to produce energy cost reductions of 20 percent or more. Aggressive programs produced savings of 35 percent and more.

In spite of this outstanding track record, there is much more that can be done—partly as a result of new buildings and systems that have come on line during a period of relatively low energy costs, and partly as a result of backsliding. Although energy management has played a key role in building codes and design standards, the lack of economic incentives has relegated it to a much lesser role in operational considerations. As a result, new buildings and systems may have the potential to be much more energy efficient, but they simply are not being operated that way.

The same holds true for older buildings and systems. Many of the efficiency gains made in the past have been lost as the result of lost interest and lost priorities. Between 1973 and 1986, overall energy use in the United States grew by 0.1 percent, primarily as a result of widespread energy conservation efforts. Between 1986 and 1995, overall energy use grew by 17 percent. While per capita energy use declined by 14 percent during the period between 1973 and 1986, by 1995, it had risen nearly to the pre-embargo level. Much of the increase can be attributed to a lack of focus on energy issues.

The same growth in energy use has taken place in facilities. In spite of new construction standards, per capita energy use in facilities has been increasing. This increase can be attributed to a lack of concern for energy issues. It also can be attributed in part to past economic practices. Many organizations, in an attempt to cut operating costs, cut back on their maintenance operations. Staffing and budgets were reduced. One of the first maintenance areas to be cut is the one with the greatest impact on energy use—preventive maintenance. We are seeing increases in energy use today as the result of those cuts made during the past ten years.

One thing is certain; the price of energy is ultimately set by supply and demand. We have been fortunate that with only a few, brief exceptions, energy supply has exceeded demand sufficiently to moderate prices. Partly as a result of these moderate prices, demand has been increasing. If the demand for energy continues to rise and reaches levels that cannot be sustained by production capacity, the price will rise significantly and rapidly. The result will be an economic crisis for organizations in general and facility managers in particular.

SUPPLY INCENTIVES FOR ENERGY MANAGEMENT

World demand for all forms of energy is increasing as a result of both population and economic growth. The world has only a finite supply of energy resources. Estimates of actual reserves vary widely, depending on who is producing the reserve estimates and how those estimates are to be used. Best-case estimates say that the world has sufficient energy to last at least to the mid-point of the twenty-first century. Worst-case estimates say that by the turn of the century we will have used one-half of the world's total supply of recoverable oil and that production will start to decline. Not only will production decline, the cost of production will rise dramatically as we will have already used the majority of the easily recoverable resources.

The actual quantity of recoverable reserves is most likely somewhere between the two estimates. However, we must accept the fact that reserves are finite and demand is increasing. Faced with increasing demand for energy, there are only two options: increase production to meet the increasing demand, or implement aggressive programs to manage energy use. In reality, managing what resources we have may be the only practical option.

The oil shortages that we experienced during the 1970s and 1980s were the result of a number of factors, including increasing world population, expanding economies, the realization that world supplies were finite, political disruptions, and policies by a number of oil producing countries to limit production. While much has changed in the world since the last oil shortage, four of these five factors remain critically important today. World population is still increasing. Energy demand is rising as countries expand their economies. There is an even greater realization of the finite nature of world energy resources, particularly by those countries that are producing those resources. Political friction, particularly in the oil-producing Middle East countries, continues. The only factor that has changed is the policy of limiting production. So far, the need for capital has prevented oil producers from limiting production. So far.

There are four factors that directly impact supply incentives for energy management: the decline in domestic oil production, the increasing level of oil imports, natural gas supplies, and the growing need for electrical generating capacity.

Domestic Oil Production

Oil drives world economies, accounting for nearly 40 percent of all primary energy use. U.S. domestic production of oil reached a peak of 9.6 million barrels per day in 1970 and has been declining ever since in spite of major exploration activities during the mid and late 1970s. By 1995, domestic oil production had declined by nearly one-third to 6.5 million barrels per day. Even Alaskan North Slope oil production is in a decline. As the Alaskan North Slope oil field continues to play out, U.S. production will decline even further.

According to the most optimistic estimates of proven national reserves, we have less than a twenty-five-year supply of oil if production levels are maintained at current levels. Even if the Alaskan wilderness is opened up to production, it would extend the life of domestic oil production by only five years. It is doubtful that many other new large-scale fields of oil will be found in the United States.

Two factors support this view. First, proven U.S. reserves are in a steady decline and have been for nearly a decade. Output from existing wells is declining and relatively few new wells are coming on line. And the decline in output is accelerating. The cheap, easy to recover oil is diminishing. What remains is more difficult and more costly to extract.

Second, the combination of relatively low prices and low success rates has resulted in fewer drilling attempts. We are currently searching for oil at far lower levels than we were ten years ago. The net result is that any further increases in U.S. oil use will have to come from additional imports.

Increasing Dependence on Imports

In 1995, the United States imported approximately 50 percent of its oil needs. By 2010, that figure is projected to rise to approximately 60 percent. With nearly two-thirds of the world's proven oil reserves residing in the Middle East, any increase in use by the U.S. will have to come from countries in the Middle East. But the U.S. will not be the only country seeking to buy additional Middle East oil. Other countries will also be forced to turn to that region for additional oil supplies. The projected ten percent increase in world demand for oil by the turn of the century will have to be met almost exclusively by Middle East supplies.

The question is, will the countries of the Middle East be able or willing to meet this increased demand? The area has a history of occupations, wars, revolutions, terrorism, and oil embargoes. Even if there are no politically induced disruptions, the Middle East may not be able to meet world demand by itself. A modest 2 percent increase per year in demand may exceed the ability or the willingness of those countries to expand their production capabilities fast enough to keep up with demand. If production capabilities do not keep up with world demand, the competition for oil will result in rapidly escalating oil prices. Current demand and production projections suggest that there is a real possibility of that occurring within the next fifteen years.

Natural Gas Supplies

The situation is somewhat better with natural gas. The United States has proven reserves in excess of 160 trillion cubic feet, and an estimated total natural gas resource of 1,000 trillion cubic feet. While it is unclear how much of the total resource can be converted into proven reserves, natural gas officials are optimistic. There are two factors that help to make natural gas an attractive energy option: environmental concerns and energy security. Natural gas burns cleaner than oil or coal, thus making it more environmentally friendly.

Natural gas is also not subject to the same curtailment and price manipulation techniques that can be used with oil, as natural gas reserves are widely distributed around the world. Supplies cannot be easily manipulated by a limited number of major producers.

While the U.S. imports natural gas each year, only 11 percent of the total use comes from imports. By the year 2010, the use of imported natural gas is projected to rise to 15 percent. Even then, we will not be placing our natural gas future at risk as the imports come from Mexico and Canada.

Natural gas use is increasing in the U.S. primarily as a result of environmental initiatives and its increased use for power generation. Increased demand along with rising process has led to more aggressive drilling programs. While many of these drilling efforts have been successful and have resulted in additions to proven reserves, there has been a steady decline in the proven reserves each year for the past ten years. The United States is simply using natural gas faster than it is being discovered.

If use rates were to be frozen at the current rate of slightly more than 20 trillion cubic feet per year, our current proven reserves would last only eight years. If the estimated resources for the country were converted to proven reserves, at our current rate of use, they would be depleted in fifty years. Depletion of natural gas reserves will have an immediate and very negative impact on facility managers.

Electrical Generating Capacity

The electric power industry in the United States is entering a new era marked by significant change. Demand for electricity is growing by 2.5 percent per year. There are fundamental changes taking place in the structure of the industry, including increased competition and industrywide deregulation. Plants are shifting their fuel mix from coal and oil toward cleaner burning natural gas. Plants are investing heavily in equipment to improve operating efficiency.

In spite of these changes, the electrical power industry is entering an era of reduced generating capacity margin. Generating capacity margin represents the national average of excess capacity that electrical producers have available. In 1995, the margin was approximately 15 percent. Assuming an annual growth rate of only 2 percent, and taking into consideration new plants that are scheduled to come on line, this margin will decrease to 10 percent by the year 2005.

Historically, the annual growth rate has averaged 2.5 percent. If future demand for electricity increases at 2.5 percent instead of the projected 2 percent, the generation capacity margin in 2005 will be only 2 percent. Industry analysts predict that if in fact the margin does decline to 2 percent, there will be widespread and lengthy electrical brownouts and blackouts. The situation would be further complicated if any plants are forced to be taken out of service due to age, economic reasons, or safety concerns.

The Impact of Supply Shortfalls

With supplies of oil, natural gas, and electricity decreasing at the same time that worldwide demand is increasing, the law of supply and demand dictates that prices will rise. How sharply and how rapidly they will increase depends on the size of the gap between supply and demand. Energy management, by reducing the demand for energy, can help to reduce the gap.

ENVIRONMENTAL INCENTIVES FOR ENERGY MANAGEMENT ———————

The energy used by buildings in the United States has a large impact on national environmental conditions. In a typical year, residential and commercial buildings account for approximately 35 percent of all energy use in the country, and result in some 470 million metric tons of carbon dioxide emissions. Although there are other environmental impacts of energy use, it is this introduction of additional carbon dioxide into the atmosphere that causes the greatest concern.

Carbon dioxide has long been recognized as one of the major greenhouse gases that may be contributing to global warming. Increases in the concentration of carbon dioxide in the atmosphere impact how effectively the earth can radiate energy into space but do not impact the rate at which it absorbs energy from the sun. The result most likely will be increased surface temperatures on earth. The best estimates of the impact of a doubling in the carbon dioxide levels from pre-industrial levels is an average world surface temperature increase of one to three degrees Celsius. Although this might not seem like much, temperatures during the last ice age were only four degrees Celsius colder than today. The impact of such an increase will be as dramatic and will most likely include more extreme weather events, a rise in sea levels sufficient to cause coastal flooding, and extensive damage to agriculture and commerce.

Since the start of the industrial revolution approximately 150 years ago, the carbon dioxide levels in the atmosphere have increased from 270 parts per million (ppm) to 350 ppm today, and are increasing at the annual rate of 1.5 ppm. As carbon dioxide accumulates in the atmosphere, it helps to trap thermal radiation from the earth's surface. Trapping the thermal radiation from the earth's surface disturbs the thermal equilibrium of the earth, resulting in increased temperatures. The question is not if this will happen, but rather, to what extent. Evidence from ice cores and tree rings suggest that the process has already started, with a 0.5 degree Celsius rise in overall mean global temperature since 1850.

The consequences of global warming represent only one environmental incentive to manage energy use. A more direct and immediate concern to facility managers is the impact that energy use has on air quality. In recent years, a large number of urban centers fail to meet U.S. ambient air quality standards. While automobiles are thought to contribute nearly one-half of the pollution found in the air at these sites, the energy used by facilities also has a significant impact on air quality. Efforts to improve air quality in these cities will directly impact both the energy generators and end users through increased costs and greater environmental restrictions.

The net result of these environmental issues will be additional pressures to reduce facility energy use through energy management.

WHERE WE ARE HEADED

The combination of economic, supply, and environmental factors will increase the need for energy management, and will increase it at an accelerating rate. As energy resources continue to diminish, their costs will increase at an accelerating rate. Temporary supply disruptions will only make matters worse. As energy costs increase, they will require more and more funds to be diverted to the operating budget. Restrictions on emissions and regulations on the use of materials commonly used in energy using systems, such as CFC-based refrigerants, will cause facility managers to invest capital in new or modified systems.

Perhaps the only effective way of countering the impact of these factors is through energy management. Energy management offers the long-term potential for improving energy efficiency and thus blunting the impact of diminishing energy resources, escalating costs, and environmental factors without disrupting the operation of the organization. By implementing good energy management today, you will help ensure that your organization is prepared for whatever lies ahead.

Chapter 4

How to Evaluate
Building Energy Performance

Passage of the Energy Conservation and Production Act of 1976 focused attention on energy performance standards for new buildings. The Act required the federal government to develop energy performance standards for new building designs. The standards were to have established a maximum allowable energy use level in a wide range of building types and climates. The performance standards, while defining maximum energy use levels, did not specify how engineers and architects were to design facilities to meet those levels. The belief was that leaving the methodology unspecified gave engineers and architects greater flexibility and allowed them to use their creativity to develop alternative designs to meet the requirements of the standards.

Reaction to the standards was initially mixed, but increasingly negative with time. While some critics contended that it was too easy to vary parameters in the energy analysis program to bring a design into compliance with the standards, the major concern was with the concept of the standards themselves. Critics argued that there simply were too many variations in building design and operating parameters for the energy use standards to be meaningful.

Today energy performance standards have largely been replaced with prescriptive standards and building codes. These standards and codes identify which energy conservation construction features must be included in new building construction, as well as minimum energy using system and component efficiencies.

While energy performance standards may have been rejected as a means of achieving energy conservation in new building design, building energy performance remains an important issue for energy managers. If facilities are to be operated as energy efficiently as possible, then energy managers must know first how efficiently their facilities are using energy, and second, how that energy use compares to other, similar facilities. By making a comparison to energy use in previous months and years within their own facilities, they will gain an understanding of how their performance is changing over time. By comparing their energy use to use in comparable facilities, they can gauge how well they are performing relative to others, and how much room there is for improvement.

Determining the energy performance of a building is not a one-time effort. Energy performance evaluations must be made on a regular basis. While the annual energy performance is calculated once a year, most facilities will benefit by determining their energy performance on a monthly basis. Larger, more complex facilities with annual energy bills in excess of $10 million will benefit from daily determinations of energy performance, particularly if systems are sub-metered.

Why is it so important to constantly evaluate energy performance of a facility? First, facilities change. Buildings change. Systems change. Functions change. Schedules change. People change. These changes will be reflected in the facility's energy use—some in positive ways, some in negative ones. Failing to take into consideration how a facility changes when evaluating its energy efficiency will lead to false conclusions about the overall energy efficiency of the facility and what can be done to improve energy efficiency.

Another reason why energy performance must be constantly evaluated is that systems within the facility change. Components wear. Systems go out of calibration. Dampers stick. Refrigerant leaks. Components fail. Systems malfunction. These changes also will be reflected in the facility's energy use. By closely tracking energy use over time, and by constantly evaluating building energy performance with respect to past performance, the cause of these changes can be identified and corrected.

Finally, it is not enough simply to check a building's energy performance against previous records of performance for the same building. Building performance must also be benchmarked against the performance of other similar facilities. Energy use may remain constant with time in a given facility, yet the energy using systems in that facility might be operating very inefficiently. For example, building energy systems that are not properly installed, have gone out of calibration, or are simply malfunctioning use significantly more energy, typically up to one-third more, than properly installed and functioning systems. Unfortunately, many facility managers are unaware of ongoing efficiency problems in their buildings. Just because systems are operating and meeting heating and cooling loads does not mean that they are operating properly. Without that comparison to comparable facilities, the energy manager would have no indication of ongoing energy efficiency problems.

HOW TO MEASURE THE PERFORMANCE OF YOUR ENERGY BUDGET

One of the difficulties in making energy performance comparisons between facilities is defining a meaningful measure. While it is simple to record daily, weekly, monthly, and annual energy use, differences between facilities, how they are operated, and the fuels that they use make it impossible to draw any meaningful conclusions just by looking at energy use. Therefore all energy use figures must first be normalized to provide a basis of comparison.

The performance measure most commonly used when evaluating building energy efficiency is the total energy use of the facility divided by its square footage, commonly known as the facilities energy budget. Expressed in Btu per square foot, the performance factor can be calculated on a daily, weekly, monthly, or annual basis, although the most commonly used measure is the Btu per square foot per year.

To calculate a facility's energy budget, each fuel used by the facility is summed for the year, the monthly, or daily, then converted to its Btu equivalent. If a specific fuel is used solely for a process, such as diesel fuel for a standby generator, it is not included in the building's energy budget. One exception would be standby generators that are used to offset peak electrical loads.

The Btu equivalents for all fuels used within the facility are then summed to determine the total energy use of the facility. This value is then divided by the total gross square footage of the facility served by those fuels. The square footage figure includes all conditioned spaces. It does not include areas such as unheated storage sheds or garages, crawl spaces, attics, or porches.

The energy budget based on Btu per square foot per year is only one performance measure. Energy managers have developed other measures in an attempt to make them more meaningful to a particular facility. One measure calculates Btu per square foot per hour that the facility is occupied to factor in varying occupancy schedules. By dividing by the number of hours that the facility is occupied, energy use in a given facility can be compared during times of varying occupancy. It also allows comparison of energy use between similar facilities that have different hours of operation.

One of the more complex performance measures calculates Btu per square foot per year corrected for heating degree days, cooling degree days, and the number of days that the facility was occupied. Each of these performance measures has its advantages and disadvantages. Data availability and how the performance data are to be used will often determine which measure is best suited to a particular facility.

The energy budget is a very effective means of evaluating the relative performance of a building. Comparing energy use for the same month in different years as well as comparing annual use in different years can provide information on what is happening to energy efficiency within a building. But the energy budget has its limitations. Energy use in a facility is complex and depends on many factors. Drawing conclusions about energy efficiency without first understanding what these factors are and how they influence energy use can be misleading.

FACTORS THAT INFLUENCE ENERGY USE AND PERFORMANCE ────────

One of the most important factors to understand when considering building energy performance is that energy use is not determined solely by the efficiency of the HVAC system, or the thermal characteristics of the building. Many other factors are involved. Studies have shown that even for identical buildings in the same location, operating on the same schedule, with identical functions, energy use will vary, and it will vary widely.

For example, during the 1970s, six school buildings were constructed in the Dallas area, all using the same design. The only significant difference in construction was that four of the buildings had central heating and air conditioning systems while two had rooftop units. All six operated the same number of hours and served approximately the same number of students. If energy use were determined solely by design and climate, all six would have similar energy use. However, annual energy use in the facilities ranged from a low of 76,536 Btu per square foot per year to a high of 212,240. Average for the six buildings was 119,586 Btu per square foot per year. While energy use in the buildings with rooftop units was slightly lower on the average than buildings with a central heating and air conditioning system, the difference was not significant enough to explain the wide variation in building energy use.

Why was there such a wide range of energy use for nearly identically constructed buildings in the same climate? The answer is that climate and construction, while important, are only two of the factors that influence energy use in buildings. There are a large number of other factors that impact building energy performance, and many have an even greater impact.

People Factors

One of the largest factors in determining building energy performance is people. People determine how the energy using systems are operated. They can set thermostats high or low. They can leave lights on or off when areas are unoccupied. They can set outside air dampers to introduce large or small quantities of fresh air. They can leave windows open while the heating or air conditioning is operating. They can choose to maintain or not maintain energy using equipment. They can bypass energy conservation features. The impact of people on building energy use is so great that it more than offsets the impact of system efficiency and building components on building energy performance. People were most likely the primary contributors to the wide range of energy use in the Dallas school buildings.

There are so many factors tied to the people who occupy and operate the facilities that they are impossible to list completely. They are, however, critical to the energy performance of a building. Identifying them will require walking through the building during occupied hours, repeating the walk through during unoccupied hours, and examining maintenance records and activities.

Building Type Factors

Building type factors are those factors related to the principal activities performed in a building. Activities to a great extent determine building energy use, both directly because of the activity and indirectly because of the type of operations that take place in the building in support of the activity.

For example, buildings that have extensive main frame computer operations have high energy use rates, typically twice that of comparable facilities without the computer operations. This energy use rate is a result of the year-round HVAC requirements of the computers, and the fact that most computer operations are twenty-four-hour-a-day, seven-day-a-week operations. Few office buildings without main frame computer facilities have similar requirements.

Figure 4.1 lists eleven building types and their average annual energy use per square foot. The data in the figure is from a survey of approximately 6,600 commercial buildings from across the United States. As can be seen in the figure, there is a wide variation in annual energy budgets based on building type, from a low of 29,000 Btu per square foot per year for religious facilities to a high of 228,500 Btu per square foot per year for health care facilities.

While the values listed in the figure can be used as a reference point in looking at the energy performance of a particular building, other factors will have an impact on actual energy use and must be taken into consideration. For example, the Association of Physical Plant Administrators (APPA) conducted a survey of higher education facility energy use. For four-year colleges, the range of annual energy use was from 33,000 to 273,000 with a national average of 139,000 Btu per square foot per year.

Occupancy Factors

Depending on the nature of the activities being performed in the facility, the building may be operated anywhere between a few hours per week to 24 hours per day. The shorter the hours of operation, the lower the annual energy use on a square foot basis as heating, cooling, and other loads related to building occupancy are reduced.

Figure 4.2 lists annual energy use per square foot for various building occupancy patters. The data were compiled from a nationwide survey of commercial facilities conducted by EIA (Energy Information Administration). The data showed the expected pattern of increasing energy use with increasing hours of operation.

Climate Factors

Climate has long been considered one of the primary determinants of energy use in a facility; the more severe the climate, the greater the energy use. Early energy conservation programs focused on limiting the effects of climate through increased insulation, lower space temperatures during the heating season, higher space temperatures during the cooling season. While some savings were achieved, most were not that significant in comparison to total building energy use. The reason for this is that while climate does influence building energy use, its impact is not as large as the combined impact of other factors.

One measure of the impact that climate has on building heating energy use is the heating degree day. The number of heating degree days for one day is defined as the number of degrees the mean temperature for that day is below 65°F. The total number of heating degree days for a heating season or a year is found by summing the values for each day.

Figure 4.1
Average Annual Energy Use by Building Type

Building Type	Energy Use (Btu/square foot per year)
Education	75,200
Food sales	181,500
Food service	206,100
Health care	228,500
Lodging	160,100
Mercantile and service	71,900
Office	101,200
Public assembly	68,000
Public order and safety	110,600
Religious worship	29,000

Source: Energy Information Administration, *Commercial Buildings Energy Consumption and Expenditures, 1992.*

Figure 4.2
Average Annual Energy Use by Operating Hours

Weekly Operating Hours	Energy Use (Btu/square foot per year)
39 or fewer	33,700
40 to 48	67,400
49 to 60	66,500
61 to 84	79,700
85 to 167	99,000
Open continuously	145,800

Source: Energy Information Administration, *Commercial Buildings Energy Consumption and Expenditures, 1992.*

While the energy used to heat and cool a building does vary with climate, other factors, such as building lighting, fan energy use, and miscellaneous electrical loads, do not. Figure 4.3 lists the average annual energy use in Btu per square foot for commercial buildings in a range of climate zones. For the buildings surveyed, annual energy use varied from 71,700 to 91,100 Btu per square foot—a variation of 27 percent, and less than the variation by a factor of nearly seven for type of building.

A common mistake made when attempting to compare energy use in buildings located in different climates or for the same building during years with different numbers of heating degree days is to assume that a 10 percent variation in the number of heating degree days results in a 10 percent variation in energy use. That relationship would hold true only if all energy use in a building were related to the number of heating degree days. Since much of a building's energy use is not related to heating, it is more likely that a 10 percent increase in the number of heating degree days would result in a 1 or 2 percent increase in building energy use.

Age Factors

Building energy use also varies with the age of the facility. Some of this variation is due to the normal aging process of energy using systems. As those systems age, their operating efficiency decreases. Scale builds up on heat transfer surfaces, dirt coats the inside of HVAC ducts, normal wear increases tolerances in control system components; all contribute to decreasing operating efficiency with age.

But there are other age-related factors that contribute to variations in energy use. Newer buildings and systems are built to tighter energy performance standards. For example, a new, high-efficiency centrifugal chiller typically uses one-half of the energy to produce the same amount of cooling as was required by a newly installed chiller ten years ago. Today's building energy codes require high energy efficiency designs that simply were not used ten or fifteen years ago.

Figure 4.4 lists the average energy use per square foot per year for buildings constructed between 1900 and 1992. From the figure it can be seen that energy use steadily increased from 1900 through 1979. Starting in 1980, building energy use has been declining. For buildings constructed between 1990 and 1992, it was down by nearly 30 percent from the 1970s level. Much of this decrease in energy use can be attributed to the use of more energy-efficient designs and systems following the rapid energy price increases of the 1970s.

Construction Factors

The EIA study examined several factors related to the building construction that impact annual energy use, including the predominant exterior wall material and the percentage of window glass in the exterior of the building. In both cases, the study found significant variations in energy use. Figure 4.5 lists annual energy use by predominant exterior wall material. Highest energy use was for buildings whose exteriors were primarily

Figure 4.3
Average Annual Energy Use by Climate

Climate	Energy Use (Btu/square foot per year)
More than 7,000 HDD and less than 2,000 CDD	84,800
5,500-7,000 HDD	91,100
4,000-5,499 HDD	78,900
Fewer than 4,000 HDD	77,200
Fewer than 4,000 HDD and more than 2,000 CDD	71,700

Nite: HDD=heating degree days; CDD=cooling degree days

Source: Energy Information Administration, *Commercial Buildings Energy Consumption and Expenditures, 1992.*

Figure 4.4
Average Annual Energy Use by Building Age

Year Constructed	Energy Use (Btu/square foot per year)
1899 and before	68,600
1900 to 1919	59,100
1920 to 1945	76,500
1946 to 1959	76,800
1960 to 1969	89,200
1970 to 1979	90,000
1980 to 1989	79,300
1990 to 1992	69,300

Source: Energy Information Administration, *Commercial Buildings Energy Consumption and Expenditures, 1992.*

Figure 4.5
Average Annual Energy Use by Exterior Wall Material

Exterior Wall Construction	Energy Use (Btu/square foot per year)
Masonry	82,700
Siding or shingles	67,600
Metal panels	68,300
Concrete panels	82,200
Window glass	96,100

Source: Energy Information Administration, *Commercial Buildings Energy Consumption and Expenditures, 1992.*

glass. Further, as shown in Figure 4.6, the greater the quantity of glass, the higher the energy use.

These are not the only construction factors that influence energy use. Energy use varies by the number of floors in the building, the shape of the building, the orientation of the building, the type of roof installed, and many other construction factors.

Other Factors

There is a wide range of other factors that influence energy use in buildings, factors that have not been studied in as much detail as the factors examined by the EIA study. For example, energy use is widely influenced by the type of energy using systems installed, how those systems are set up and operated, and how well those systems are maintained. It would be interesting to see the results of a study that examined maintenance expenditures and energy use in facilities.

Another factor that has a major influence on building energy use is how aggressively energy management is pursued. Energy management systems have been widely used in facilities of nearly any size with very positive results—particularly in those facilities where energy managers work with the systems on a daily basis to minimize energy use. Other energy management activities also produce very effective results.

The variations in energy use in the Dallas school buildings and the buildings in the EIA study demonstrate that there are many factors that contribute to a building's overall energy use—some obvious, some not so obvious. It is an important point for energy managers: having energy-efficient equipment is not sufficient for good energy performance. Other factors must be examined and taken into consideration.

Equally important, the examples demonstrate that one cannot simply compare energy use in a particular building to energy use in similar buildings without first taking into consideration the differences in use and operation between the two buildings. Attempting to compare buildings that are not similar will give an inaccurate measure of the energy efficiency of a given facility.

HOW TO EFFECTIVELY USE ENERGY PERFORMANCE DATA

In spite of its limitations, the building energy budget is still an effective tool for evaluating relative building energy performance and for identifying potential energy efficiency problems as they develop in a particular facility, as long as it is used properly. It helps an energy manager answer two key questions: How effectively is your building operating in relation to other similar buildings? Is the building operating as effectively as possible? Proper use requires that one understand how different factors influence building energy performance, and that one must take these into consideration when comparing energy use in different but similar buildings.

If the facility being evaluated is relatively simple and contains routine functions that do not change with time, energy budget comparisons can be made to previous years without any modifications. Weather variations in terms of heating degree days and

Figure 4.6
Average Annual Energy Use by Percent Window Glass

Percent Window Glass	Energy Use (Btu/square foot per year)
25 or less	76,000
26 to 50	94,300
51 to 75	97,300
76 to 100	107,100

Source: Energy Information, *Commercial Buildings Energy Consumption and Expenditures, 1992.*

cooling degree days should be noted but not directly factored into the calculation. After several years' worth of data has been collected, a pattern will develop that will show how much variation in energy use can be expected for a given variation in weather.

Similarly, comparisons can be made to other buildings as long as the functions performed in those buildings are similar and the hours of operation also are similar. Again, note variations in the number of heating degree days and cooling degree days but do not factor them into the calculation of the building's energy budget.

More complex facilities will require a more detailed evaluation to normalize the values for variations between and within buildings. If the occupancy and functions performed with a building do not change over time, then energy budgets can be compared from one year to another without corrections. Again, note any variations in seasonal heating and cooling degree days. Comparisons to other facilities will be more difficult as there will most likely be differences in functions performed in the buildings as well as differences in occupancy patterns.

In both simple and complex facilities, look for changes in the calculated energy budget over time. If those changes cannot be explained by a functional change in the building or by unusual weather, then it is time to look for causes within the building systems. A slow, steady decline in energy performance is generally an indication that additional routine and preventive maintenance are needed to bring the energy using systems back to like-new condition and operating efficiency. A sudden, sharp drop in building energy performance that cannot be explained by changes in weather or use of the building are often signs that something has failed in an energy using system.

Finally, the more data available, the more useful the energy budget will be in evaluating building energy performance. Monthly, weekly, and even daily data are particularly valuable when trying to identify specific problems in energy using systems, especially in large facilities.

Chapter 5

How to Analyze Utility Bills
in Order to Reduce Energy Costs

When organizations establish energy management programs, most start with an examination of energy using systems. Many perform elaborate energy audits that identify where energy is being used by systems in the facility and how much energy is being used by those systems. Complete audits go on to identify what the energy conservation options are for that facility, how much they will cost, and how much energy and money they will save. While each of these tasks is an essential element in a comprehensive energy management program, they should be performed only after the energy manager understands how much the facility is paying for energy, what the elements of the utility bill are, and what components of the utility bill can be controlled by the facility.

Why is it important to approach energy management from the perspective of the bill? With the exception of times when a facility is facing the threat of energy supply interruptions or curtailments, the primary motivation of energy management programs is and always has been saving money. In order to gain the best rate of return on investments in energy conservation, it is necessary to implement actions that will have the most positive impact on the organization's energy bill. And in order to identify those actions, it is necessary that the energy manager first understand what the facility is being billed for and why, and what elements can be controlled. That understanding begins with a detailed analysis of the way in which the utility company bills for energy.

Analyzing a utility bill and the utility's rate structure in detail is not overkill. It is essential to gaining an understanding of what can be done about energy costs. The information that can be gained from that understanding will help in designing the most effective program for reducing energy costs. Since utilities have a wide range of rate structures in effect for commercial, industrial, and institutional customers, and rate structures vary widely among utilities, the analysis must be specific to the facility and must be based on the rate structure that is in effect. Attempting to apply a generalized rate analysis will lead to false assumptions about energy costs and the cost savings of potential energy conservation measures. For example, the cost benefit of a night shutdown of HVAC systems will be largely over estimated if average electricity costs are used when the facility is on a time-of-day rate structure. Under time-of-day rate structures, off-peak energy use typically costs one-third to one-fourth as much as the facility's average electricity cost.

In spite of the importance of understanding utility bills, few energy or facility managers take the time to review them in detail, or to learn more about them beyond what their total cost is and which way the monthly and annual charges are headed. Most view rate schedules as being complicated and difficult to understand. Their interest extends only to reducing the bottom line on the utility bill; not a detailed understanding of each line item.

Rate structures are not overly complex, although there are a number of individual components that must be examined. Energy use charges, transmission charges, demand charges, fuel adjustment charges, minimum charges, and ratchet clauses are all among the more commonly used components of rate structures. Knowing what they are and how much control you have over each is the first step toward getting energy costs under control.

ELECTRICAL BILL COMPONENTS

kWh Charges

Electrical energy use is billed for the total kilowatt hours (kWh) of electricity used during the billing period. In some smaller facilities, it may be billed at a flat-rate charge per kWh, regardless of the time when it was used, the season when it was used, or the total volume of use. More often, the billing rate varies significantly with all three variables. For example, summer season on-peak cost per kWh of use can be four or five times that for summer off-peak use, and at least twice the cost of winter on-peak costs.

While some utilities charge a flat rate per kWh of electricity used, regardless of the quantity used, most utilities follow a declining block-rate structure. The net effect of declining block rates is that the more electricity used by a customer, the lower the unit cost is for that electricity. Under declining block rates, the customer is charged a certain rate for the first quantity or block of electricity used. Each additional block of electricity use is charged at a lower rate. Most rate structures use three to five separate block rates. The motivation behind block rates is not to reward customers who use large volumes of electricity, rather, it is a recognition that the cost of providing service to customers decreases with increasing use.

Electrical use charges vary widely from utility to utility, in part due to the type of generation system used by the utility. Rates are lowest in the Pacific Northwest where much of the electrical supply comes from hydroelectric plants, and highest in the New England states where nuclear and oil generation plants are more widely used.

Demand Charges

Electrical demand charges are based on the facility's largest demand for electricity measured over a set interval during the billing period. The typical measuring interval is 15 or 30 minutes. Electrical demand, measured in kilowatts (kW), is the way that the utilities attempt to recover their costs for installing sufficient generating capacity to meet their customers' needs. If the demand for electricity were the same throughout the year and throughout the day, then the generating equipment would be in use most of the time. However, since demand varies with time of day and season, utilities must maintain the peak capacity required by their customers even though that capacity may be needed for only a few days per year. Demand charges help to recover these costs and help to encourage customers to limit their demand for electricity.

Demand charges vary widely by utility, falling somewhere between a high of $20 per kW and a low of $3 per kW. Many demand schedules are structured as block rates, similar to rate schedules. Higher blocks of demand are charged at decreasing rates. For customers with rate schedules that have high demand charge rates, the demand portion of the electrical bill often equals or exceeds the energy use portion of the bill during summer months.

Electrical Demand Ratchet Clauses

A ratchet clause is added to many utility demand charges as an additional way of recovering utility generation capacity expenses and to provide further incentive for customers to reduce their peak demand. The ratchet clause allows the utility to bill for peak demand at the actual peak demand level achieved during that month, or at 60 to 70 percent of the highest peak demand that occurred during the previous eleven months, whichever is greater. Customers who have a very high peak demand during summer months, yet a low peak demand during winter months, typically end up paying the ratcheted 60 or 70 percent peak demand, a significant cost increase over actual demand.

Power Factor Charges

The current that is required to drive induction motors, fluorescent and HID lighting systems, induction furnaces, and some other electrical equipment consists of two components—the power-producing current and the magnetizing current. Power- producing current, measured in kW, is converted into useful work by the equipment and is billed for by the utility in the energy use charges for the facility. Magnetizing current, measured in kilovars (kvar), is the current that produces the magnetic flux required for the operation of induction devices. The total or apparent power in the circuit is the sum of the power-producing current and the magnetizing current. Power factor is the ratio of the work-producing current to the total current drawn from the utility. In most facilities, except industrial applications with a large number of induction motors, power factor typically ranges between 0.90 and 0.95.

Miscellaneous Charges

There are a number of additional charges on most electrical bills that are fixed and are not controllable by the customer. The two most common are administrative or customer charges and fuel adjustment charges.

Customer charges are implemented by utilities to recover their fixed costs of providing service to the customer, including providing a meter, reading the meter, processing the bill, and providing customer support. Customer charges are set by the utility and remain the same regardless of energy use. As such, they are not impacted by energy management efforts.

Fuel adjustment charges are a means that the utility has to pass on to its customers any cost changes in the fuel they use to generate electricity. When rates are set in rate schedules, the average cost of the fuel used to generate the electricity is assumed by the utility. Often, particularly when fuel prices are rapidly changing, the estimated average cost is inaccurate. The fuel adjustment charge is used to correct the estimated costs that were used to calculate the rate schedule for the actual fuel costs. Since the original rates were estimated and could be higher or lower than the actual costs, the fuel adjustment charge could be a charge or a credit.

Fuel adjustment charges, like customer charges, are set by the utility. However, since they represent an additional cost per kWh of electricity used, their total cost to the customer depends on actual energy use during the billing period. Energy management programs that reduce overall energy use in a facility will also reduce the impact of the fuel adjustment charge.

ELECTRICAL RATE STRUCTURES

The simplest rate schedule for electricity is the general service rate. Customers who are billed on a general service rate schedule typically pay a flat rate for electricity use based on the total kWh of electricity used during the billing period, a demand charge based on the peak kilovolt amps (kVa) supplied during the billing period, and a number of other miscellaneous charges, including fuel adjustment charges. Once widely used for commercial and small industrial customers, this rate schedule has all but been replaced with more complex rate schedules, such as the time-of-day schedule.

A modified form of the general service rate adds a third component, a load factor. Not to be confused with electrical demand, the load factor is a measure of the average use of electricity with respect to the maximum use during that billing period. For example, if a customer's use of electricity were constant over the billing period, the load factor would be 100 percent. More likely, the use of electricity will vary during the billing period, resulting in a lower load factor. The lower the load factor, the greater the charge for service.

For customers who are operating under a general service rate, there are two ways to reduce the cost for electrical energy: energy use reduction and demand limiting. While the impact of demand limiting will be restricted to the period of time when the demand for electricity in the facility is the greatest, savings from energy use reduction occur throughout the billing period. Any reduction in kWh use, regardless of when that reduction takes place, will result in lower energy costs. With the same rate in effect regardless of the time of day, the savings produced by any reduction in kWh will be the same no matter when that reduction takes place. Typical energy management activities that will benefit customers on a general service rate are unoccupied hours equipment shutdown, primary heating and cooling equipment efficiency improvement, and demand limiting.

The most common rate schedule in use today is the time-of-day (TOD) structure. TOD rate structures eliminate the flat rate pricing of electricity, replacing it with a pricing schedule that varies with the time of the day, the day of the week, and the season of the year. They were developed by utilities as a way to reduce the need for peaking stations. Peaking stations, typically oil or gas fired, are more expensive to operate than their base load plants, thus increasing the cost of generating electricity. TOD rates are a way to pass on the increased cost of generating electricity during peak periods, and to encourage customers to flatten out their demand for electricity.

The typical TOD rate schedule includes three periods: on-peak, intermediate peak, and off-peak. On-peak typically runs from noon to 8:00 P.M. on weekdays. The intermediate peak period typically runs from 8:00 A.M. to noon, and 8:00 P.M. to midnight, also on weekdays. Off-peak includes all other times, including holidays. The rate schedule also divides the calendar into two seasons, summer and winter months. Summer months typically include the months of June through October.

Each of these six periods identified in the rate structure, three summer and three winter, has its own cost per kWh of electricity used. Demand charges are determined by the peak demand during the peak period, with two rates in effect: one for summer and one for winter.

What makes the TOD rates structure particularly effective for the utilities is the variation in rates between on-peak, intermediate peak, and off-peak periods. While actual rates vary, in general on-peak rates are approximately one-and-one-half times the rates during intermediate peak periods, and more than three times the rates during off-peak periods. Similar variations exist in demand charges, with summer month demand rates being more than double winter ones.

TOD rates have a very strong impact on the effectiveness of energy conservation measures. Under time-of-day rates, energy conservation efforts must address both the energy use and the demand portion of the bill. While any reduction in kWh use, regardless of when that reduction takes place, will result in lower energy costs, the TOD rate structure increases the cost effectiveness of measures that impact energy use during on-peak hours, while decreasing the cost effectiveness of measures that impact off-peak energy use. The cost effectiveness of measures that impact on-peak energy use is increased further by the savings in demand charges those measures produce. Additionally, measures that decrease use during the summer months are more cost effective than measures that reduce energy use during winter months.

ADDITIONAL RATE FACTORS ————————————————

Power Factor

There are two major methods used through the billing process by which utility companies encourage customers to maintain a high power factor. Both impact the demand portion of the bill—one indirectly, one directly. Under the indirect method, the utility sets a minimum value for power factor, typically 85 percent. If the facility's leading or lagging power factor falls below this value, the customer's demand charges are multiplied by a factor, such as 1.1. Under the direct method, the utility bills the customer for power factor based on kVa, not kW. By using kVa, the power factor is automatically factored into the billing process.

The actions that energy managers must take in order to minimize charges resulting from low power factor depends on the billing method used by the utility. If the rate structure specifies a minimum value with a penalty clause, then it is cost-effective for the energy manager to implement a power factor correction program to ensure that the facility's power factor remains above the specified minimum value. Once that value is achieved, no further correction is justified in terms of energy conservation and cost savings. Other factors, such as distribution capacity, may come into play, but not energy conservation.

If the rate structure includes power factor in the peak demand billing, the situation is more complicated. Increasing the facility's power factor from its current value to one closer to 1.0 will produce savings. Unlike rate structures that use the power factor penalty clause, any increase in power factor will produce savings. How much savings depends on the facility's peak demand, the demand charges, and the amount of improvement. ***Rule of Thumb:*** It is not cost-effective in most applications to increase power factor beyond 0.95.

High Voltage Service

There are two different riders that can be added to commercial, industrial, and institutional rate structures if the customer can accept service at a higher than normal voltage: primary service and high voltage services. The primary service rider applies for customers who are able to accept service from the utility at voltages of 13 kilovolts (kV) or higher, directly from the utility's distribution system. Under this rider, the customer is responsible for providing and maintaining all necessary transformers, switchgear, disconnects, regulators, and protective equipment. In return, the utility discounts the total electric bill by five percent, including use, demand, and fuel adjustment charges. The high voltage service rider is similar to the primary service rider, except that it applies to customers who can accept service at voltages of 69 kV or higher. Under this rider, the customer receives a 10 percent discount on the total electric bill.

The economic advantage to the customer of reduced electrical rates is offset somewhat by the need and cost of maintaining the necessary substation equipment. To be cost-effective, the customer must be able to use service at a high voltage, must have sufficient electrical use and demand, and must have the ability to maintain or contract for the maintenance of the required equipment.

Curtailable Service

To help reduce high peak demand, many utilities have incorporated a curtailable service rider in their rate schedules. Under this rider, the customer agrees to a specified maximum demand load. When the utility notifies the customer of the need to curtail loads, the customer must reduce demand for electricity to the agreed upon demand, and keep the demand at or below that level until the curtailment period is ended.

Curtailment periods occur only during the summer months, typically June through October, and during the peak demand hours between noon and 8:00 P.M., excluding holidays and weekends. Load curtailments are limited in duration, generally six hours or less, and are limited in the number of occurrences during a given year. Customers are given a warning of at least 30 minutes during which they must decrease their demand to the set level.

The benefit for the customer is a credit for each kW of reduced demand during the curtailment period, typically between two and five dollars per kW of demand. There is, however, a penalty. If customers fail to reduce their demand to the agreed upon level, the utility imposes a charged penalty for each kW of demand that the customer failed to reduce. The charge per kW is typically several times the credit offered.

Curtailable service is best suited for facilities that have the ability to reduce their electrical load easily. Most users of this rate schedule ride are large facilities with computer-based energy management systems with a large number of controlled electrical loads that can be readily shed.

NATURAL GAS RATE STRUCTURES

Natural gas rate structures are simpler than those for electricity. Unlike electricity, natural gas rate schedules do not include peak demand charges. As a result, the majority of the costs in the natural gas bill are found in the use charge and the fuel adjustment charge. Use charges vary widely throughout the country, with the lowest rates found in the gas producing areas of Oklahoma, Texas, and Louisiana.

There are two ways that natural gas use charges are calculated: flat and block rates. Under a flat rate schedule, the customer is charged a fixed rate per 100 cubic feet (CCF) of natural gas that is used. The rate does not vary with volume of use. The use of flat rate structures is limited primarily to smaller customers.

Block rate structures for natural gas are very similar to those used by electrical utilities, with the unit cost of natural gas decreasing as use increases. These rate structures, like their electrical counterparts, are designed to reflect that the cost of providing service to customers decreases with increasing use.

Most natural gas suppliers use a single rate year round, but a few have moved to a summer/winter rate schedule. Under this rate schedule, winter months extend from November through April, with winter rates running approximately 5 to 10 percent higher than summer rates.

While suppliers of electricity experience their peak loads during summer months, suppliers of natural gas experience their peak loads during the winter months, typically December, January, and February. These peak loads can exceed the company's ability to supply natural gas. To help them meet this demand, natural gas companies use customer priority classes. Customers are divided into as many as four classes based on their priority for using natural gas. Customers with the highest priority are not curtailed during periods of high demand except when absolutely necessary. Customers with the lowest priority have their service interrupted whenever a supply shortage exists.

To encourage customers to sign up for lower priority rate schedules, natural gas companies offer significantly reduced rates for such service. While actual rates and discounts vary with the supplier, rates typically vary by as much as 15 to 20 percent between the highest and lowest priority. For energy managers, this means that significant savings can be achieved by moving from a high priority rate schedule to a lower one, provided that the facility can go without natural gas during curtailment periods or if the facility has an alternate fuel available for use.

The impact of energy conservation efforts on natural gas costs is simpler than with electricity costs. Under all rate structures, the savings produced by reducing use are the same regardless of time of day. Seasonal differences exist only for those customers whose rates are structured into a summer/winter rate schedule.

Natural gas rate structures, particularly those with different summer and winter rates, may offer energy managers a way of reducing summer electrical demand through the use of natural gas–driven absorption chillers. Some natural gas utilities offer additional rate incentives for customers to use natural gas–fired cooling systems. To determine if the system is economically justified, energy managers will have to work with their local utility companies.

USING THE INFORMATION

Utility rate structures offer energy managers many opportunities to reduce energy costs by working within the framework of the rate structure. They help to identify areas where energy management efforts will have the most impact and will provide the greatest rate of return. They offer energy managers options when looking for the most favorable rate structure for their facility. To use them effectively, all that is required is a thorough understanding of what they are and how they are applied.

Section 2

Mechanical Systems Operation
and Maintenance Practices
that Will Reduce Energy Use

Chapter 6

Chiller Systems

This chapter presents guidelines for the maintenance of chiller systems. These guidelines are generalized for typical systems and components operating in what would be considered to be routine applications, serving normal building or process loads. You may have to modify these guidelines to match the specific maintenance requirements of your particular application. If modification is required, the best source for additional information is the equipment manufacturer. Nearly every manufacturer of chiller system components has published their recommendations on the level of maintenance required for their equipment. They will assist in developing a maintenance program specifically tailored to your needs.

For the purpose of this chapter, chiller systems are defined to include a building's primary cooling equipment: the central chiller, chilled and condenser water piping, chilled and condenser water pumps, and the cooling tower. Discussion of the energy impact of maintenance will be limited to these components. Other components of a facility's air conditioning system will be covered in subsequent chapters.

CHILLER SYSTEM ENERGY MAINTENANCE CONSIDERATIONS

Building chiller systems are big-ticket items for facilities. They are expensive to purchase, operate, maintain, and replace. They are equally as costly in terms of lost time and productivity should they fail.

In nearly every case, the load placed on the electrical system by the chiller is the single largest electrical load in the facility. When the energy use of the chiller's supporting pumps and cooling tower is factored in, chiller systems are very big energy users. In order for chiller systems to operate efficiently, they must be properly maintained.

Of all building systems, perhaps the one that has the greatest return on maintenance investment is the building's chiller system. Comprehensive chiller system maintenance will increase equipment life, increase system reliability, decrease energy use, decrease the use of refrigerants, decrease the cost of breakdown maintenance, and decrease the costs of downtime. ***Rule of Thumb:*** For every maintenance dollar spent, ten dollars are saved in operating costs and breakdown repairs.

With so much riding on proper maintenance of the chiller system, one would expect that all chiller systems are well maintained. Unfortunately, few are. In practice, you are equally as likely to find an attitude of, "if it ain't broke, don't fix it," as you are to find one that values comprehensive predictive and preventive maintenance. Unfortunately, even those who feel they are doing a good job of maintenance are often failing to perform many of the tasks necessary to keep systems operating efficiently. Evidence for this is found in surveys of facilities that examined the condition of chillers and cooling towers. In one study of cooling tower condition, nearly 70 percent of the towers were found to have defects sufficient to decrease the effective capacity of the tower by at least 10 percent. Nearly 50 percent of the towers could not operate at more than 80 percent rated capacity.

Decreased tower capacity translates into higher return water temperatures, which costs facility managers through increased energy use. Each degree increase in tower return water temperature results in a 2 to 3 percent increase in chiller energy require-ments. For example, a 500-ton chiller system operating with a condenser water return temperature three degrees above design uses between 8 and 9 percent more energy to accomplish the same cooling. For an average electric rate of $0.09 per kWh (including demand), this translates into an additional expense of $9.00 per hour, $90.00 per ten-hour day, $2,070 per month, and $12,420 per six-month air conditioning season.

Why, if these components and systems are so important to building energy use, are they not properly maintained? Some facility managers simply do not understand the impact that poor maintenance has on chiller systems. Others may perform only a cursory inspection, not realizing the level of detail and attention required. Lack of personnel is a major cause of neglect, particularly for those organizations that are in a breakdown maintenance mode rather than one of preventive maintenance. Finally, lack of funding is frequently an issue. Chiller systems are not income producing in the eyes of the organization and therefore it is difficult to justify spending money on them, particularly when they are "working."

To be effective in breaking out of this mold, you must focus attention on chiller systems. You must show those who control maintenance funding how spending money on maintaining chiller system components will save both maintenance funds and energy. It is your responsibility to build the case for chiller system maintenance.

FOUR TYPES OF BUILDING CHILLERS AND THEIR ENERGY REQUIREMENTS

There are four major types of chillers used in facilities for air conditioning: centrifugal, reciprocating, rotary, and absorption. Each has its own characteristics, advantages, disadvantages, and maintenance requirements. Their energy efficiency varies by chiller size and loading as well as by chiller type.

Chiller manufacturers can provide data on the rated energy efficiency of their units. For electrical driven units, energy efficiency is rated in kW of electricity use per ton of cooling produced. For absorption units, energy efficiency is rated in Btu of heat input per ton of cooling produced. In comparing the actual performance of a particular chiller to the manufacturer's rating, it is important that the chiller be operated under the same conditions, typically full load, 44 degree chilled water supply, and 85 degree condenser water temperature.

Centrifugal Chillers

Centrifugal chillers are variable volume displacement units that use rotating impellers to compress the refrigerant vapor. Cooling capacity is regulated through the use of inlet vanes to restrict the flow of refrigerant to the impeller. Although most centrifugal chillers in use today are driven by an electric motor, they can be powered by natural gas or diesel engines as well as by steam turbines.

Centrifugal chillers are popular choices for facilities with medium to large cooling loads. Available in capacities from 90 to 10,000 tons, they can be used to cool a wide range of loads. Most of the installed units range from 150 to 500 tons.

Operating efficiencies of centrifugal chillers have been increasing during the past years. Ten years ago, new centrifugals had full-load operating efficiencies in the range of 0.70 to 0.85 kW/ton. Changes in chiller designs have improved these values to the point where most new centrifugal chillers are rated between 0.50 to 0.60 kW/ton at full load.

One of the most significant drawbacks of the centrifugal chiller is its part-load efficiency. In order to respond to decreasing building loads, the chiller partially closes its inlet vanes, throttling the flow of refrigerant. Inlet vane controls provide good regulation of chiller output, but at a cost of decreased operating efficiency. For example, a chiller with a full-load efficiency rating of 0.60 kW/ton might require as much as 0.80 kW/ton at half load, and even more at lower cooling loads.

Centrifugal chillers offer the advantage of having fairly low maintenance requirements. Their maintenance, however, is critical. Ignoring maintenance on a centrifugal chiller will lead to poor performance and early failure requiring costly repairs and extensive downtime.

Reciprocating Chillers

Reciprocating chillers are positive displacement compressors that use one or more pistons to compress a refrigerant gas. Nearly all are driven by electric motors. Cooling capacity is controlled by regulating the refrigerant gas flow to the compressor and by adding or removing compressors in a multiple compressor system.

Reciprocating compressors are popular choices in smaller facilities with cooling loads of 100 tons or less. While units are available in sizes ranging from 10 to 200 tons, the size and efficiency of the units limit their application to loads of 100 tons or less.

Not as efficient as centrifugal chillers, most reciprocating chillers have full-load operating efficiencies in the range of 0.84 to 1.00 kW/ton when new.

Reciprocating chillers offer you several advantages. For 100-ton or less units, they are lower in cost than other chiller types. They operate with a higher condensing temperature, thus making them well suited for applications requiring the use of air-cooled condensers. Varying cooling loads are readily matched by installing multiple units. The use of multiple units also allows operators to stage their operation for part-load conditions without sacrificing efficiency.

There are two significant drawbacks to their application: maintenance and noise. Reciprocating chillers have a high level of maintenance requirements in comparison to other chiller types, and their operation generates a higher level of noise and vibration than other chiller types.

Rotary Chillers

Rotary chillers are positive displacement compressors that use two machined rotors to compress refrigerant gas between their lobes. All are electric motor driven. Cooling capacity is controlled by regulating the refrigerant gas flow to the compressor.

The use of rotary chillers is limited to applications where size and weight restrictions prevent the installation of other chiller types, particularly centrifugal chillers. Units are available in sizes ranging from 20 to 2,000 tons, with most applications falling in the range of 175 to 750 tons. Their full-load efficiencies range from 0.70 to 0.80 kW/ton resulting in a chiller that is more efficient than a reciprocating chiller, but less efficient than a centrifugal.

In spite of their high initial costs, rotary chillers offer several advantages. Their compact size and low relative weight make them suitable for installations where other chiller types could not be installed. Their small number of moving parts results in lower maintenance requirements and quieter, low-vibration operation.

Absorption Chillers

Unlike other chiller types that are driven by mechanical energy, absorption chillers use heat as their energy source. There are two types of absorption chillers: direct and indirect fired units. Direct fired absorbers burn natural gas or other comparable fuel to generate the heat required to drive the unit. Indirect fired absorbers use low pressure steam, hot water, or waste heat as their energy source. The most commonly used refrigerants are water and ammonia.

Absorption chillers range in capacity from 100 to 5,000 tons, with most units falling in the 300- to 500-ton range. Thermal efficiencies of these units range between 11,000 and 19,000 Btu of heat input per ton-hour of cooling produced.

Although the thermal efficiency of absorption chillers is less than for centrifugal chillers, they are frequently used as an alternative, particularly where facilities are faced with high demand charges for electricity, or where there is not sufficient electrical capacity to power an electrically driven chiller. Indirect fired absorbers also offer the advantage of being able to be powered by a wide range of heat sources, including low pressure steam, hot water, solar energy, and waste heat. Absorbers are particularly well suited for facilities that have a readily available source of waste heat that must be disposed of.

The primary disadvantages of absorption chillers are their relatively high maintenance requirements and their high initial cost. Lithium bromide, the most commonly used absorbent, is a salt and is extremely corrosive to steel in the presence of oxygen, so units must protected through the use of corrosion inhibitors and must be kept airtight.

HOW TO MAINTAIN BUILDING CHILLERS FOR ENERGY EFFICIENCY

There is nothing better for improving the performance, reliability, and life of building chillers than a comprehensive scheduled maintenance program. Programs are most effective when they are specifically tailored to the particular application, balancing maintenance efforts with the reliability and chiller efficiency wanted. In spite of these variations, all chiller maintenance programs will have three common elements: maintenance logs, scheduled inspections, and scheduled maintenance activities.

One of the most effective tools for chiller maintenance is the maintenance log. The logging and analyzing of data on a regular basis keeps operators informed as to how well the chiller is performing. Data recorded when the chiller is operating properly provides a baseline of information for comparison when problems develop. The logs give operating engineers and maintenance personnel an operating track record for the chiller. Reviews of the data will help in the identification of trends in chiller performance—trends that develop slowly enough to not be noticed in normal day-to-day operations. To be effective, the logs must be completed every day and must be reviewed on a regular basis. Use the log in Figure 6.1 to record typical chiller operating data. New chillers, with their microprocessor-based control panels, can simplify the process of collecting the data by automatically capturing the necessary data and saving it in a file for future reference.

Scheduled inspections for chillers are generally conducted on a daily, weekly, monthly, and annual basis. Some large chiller plants conduct an inspection of chiller operations at least once per shift. The purpose of conducting inspections so frequently is to detect minor problems before they escalate into major ones that result in reduced performance or major repair bills. Many of the inspections can be conducted while the chiller is operating, taking only a few minutes to complete. Other inspections require that the chiller be shut down and in some cases, partially disassembled.

Scheduled maintenance activities also are conducted on a daily, weekly, monthly, and annual basis. Additional activities may be performed every two or three years as required by the condition of the chiller and the environment in which it operates. As with scheduled inspections, scheduled maintenance activities are designed to maintain the operating efficiency of the chiller as well as to correct minor problems before they

Figure 6.1
Chiller Log

1. Building/facility: _____

2. Room #: _____   3. Chiller #: _____

4. Chiller type: _____

Date:					
Time:					
Drive motor voltage:					
Drive motor current:					
Condenser water supply temp:					
Condenser water return temp:					
Chilled water supply temp:					
Chilled water return temp:					
Evaporator pressure:					
Condenser pressure:					
Refrigerant liquid temp, condenser:					
Refrigerant liquid temp, evaporator:					
Oil temp:					
Oil pressure:					
Oil level:					
Oil added:					

escalate into more significant and costly ones. Some activities can be performed while the chiller is operating while others require a scheduled shutdown.

Before establishing a schedule for chiller inspection and maintenance, it is important to understand what factors impact chiller operation, and what can be done to control them. There are three major factors that negatively impact chiller performance: fouling, corrosion, and scale. Controlling all three factors is particularly important to energy managers as each can significantly decrease the operating efficiency of a chiller. Each of these factors can be readily controlled through a regular schedule of inspections and maintenance.

How Fouling Works in Chiller Systems

Fouling is the process by which solids that are suspended in water circulating through the chiller drop out of suspension and slowly accumulate on the chiller's heat transfer surfaces. As the solids build up on the surfaces, they reduce the efficiency of the transfer of heat from the refrigerant to the circulating chilled and condenser water. Additionally, the scale prevents corrosion inhibitors that are added to the circulating water from reaching the metal of the heat exchangers. If fouling is allowed to progress uncorrected, it will decrease the operating capacity of the chiller, increase the energy required to produce a certain level of cooling, increase stresses on chiller components due to elevated operating temperatures, and result in extensive corrosion damage or plugging of passages in chiller heat exchangers. Unfortunately, unless close attention is paid to operating conditions or unless the chiller operates at full load for a significant period of time, fouling usually goes unnoticed. Its effects are so hidden that many maintenance personnel simply do not recognize fouling as a serious problem.

But fouling is a serious problem, particularly when it comes to chiller operating efficiency. Even a very small buildup on surfaces increases the resistance to heat transfer within the chiller. This resistance to heat transfer in a chiller is measured by its fouling factor. The higher the fouling factor, the greater the resistance to heat transfer and the lower the efficiency of the chiller. When a chiller is new and its heat transfer surfaces clean, its fouling factor is typically 0.0002 or less. Most manufacturers recommend taking corrective action when the fouling factor increases to 0.005.

After one year of operation, even with good maintenance practices and a thorough water treatment program, the fouling factor will increase to approximately 0.001, resulting in the chiller's operating with elevated condenser temperatures and pressures. Without good maintenance and a comprehensive water treatment program, it will increase even further.

It is not uncommon to find chillers that appear to be in good operating condition that have a fouling factor of 0.0025 or higher. A fouling factor of 0.0025 causes sufficient resistance to heat transfer to increase chiller condensation temperatures by four or five degrees, and condensation pressure by five to eight psi. This rise translates into an increase in compressor energy requirements of approximately 25 percent to produce the same cooling. Figure 6.2 shows the relationship between chiller fouling factor and chiller efficiency.

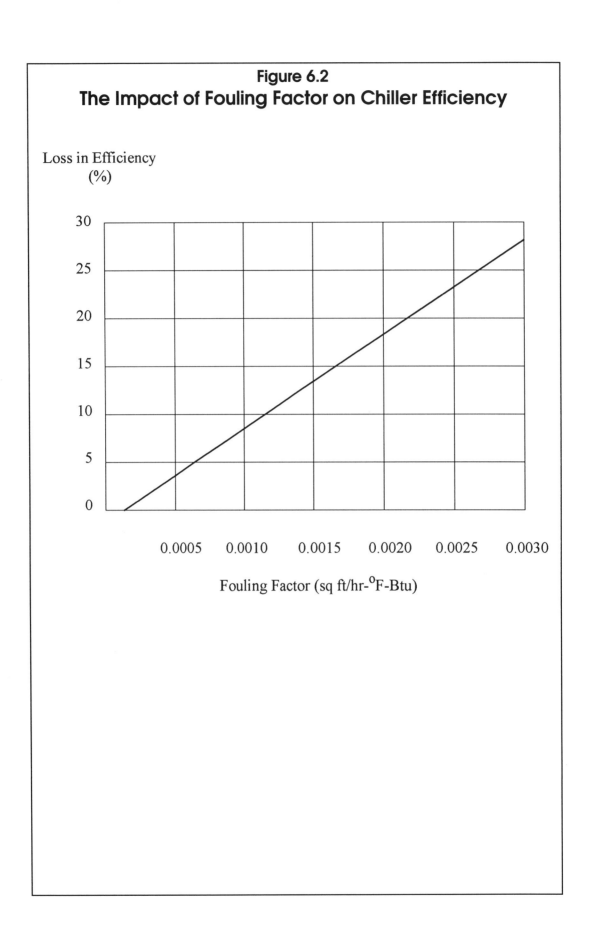

Figure 6.2
The Impact of Fouling Factor on Chiller Efficiency

There are three sources for the solids that cause fouling: solids contained in the water used by the system, solids introduced by the system itself, and solids introduced into the system by the cooling tower. All water contains at least some level of insoluble substances. As water is lost from the system through evaporation and drift at the cooling tower, the concentration of these solids in the condenser water system increases. Makeup water, used to replace water lost through the tower, introduces additional solids into the system.

The second source for solids that cause fouling is the system itself. Metallic corrosion of piping and erosion of surfaces introduces solids that become suspended in the water system until they precipitate out as the result of chemical reaction, generally on heat transfer surfaces.

The third and by far the largest source for solids that cause fouling is the cooling tower. Cooling towers are very effective scrubbers of the air. Microscopic organisms and particulate matter, including leaves, dust, and silt, are effectively removed from the air by the operation of the tower, and enter the condenser water system.

The introduction of large and microscopic organisms into the condenser water system can lead to biological fouling of the chiller's heat exchanger. Large organisms, such as weeds and floating debris, can readily block chiller tubes if not removed. Microscopic organisms, including algae, fungi, and bacteria, accumulate in the condenser water system and form slime that can interfere with proper water distribution, restrict water flow in chiller tubes, and accelerate corrosion of metal surfaces. For example, a layer of slime deposits only one one-hundredth of an inch thick on a chiller's condenser tubes raises the fouling factor by nearly 0.001 and causes the chiller to use approximately 10 percent more energy to produce the same amount of cooling.

How to Reduce Fouling

Fouling in chillers is readily controlled through a combination of operational and maintenance practices. One operational practice that can be implemented is to ensure that water flow rates in chiller systems are within the ranges specified by the chiller manufacturer. Fouling rates increase as flow rates decrease through chiller heat exchangers, particularly in those areas where the velocity of the water drops off sharply. Keeping condenser and chilled water flow rates within the recommended range will decrease the chances of fouling of chiller tubes and end plates.

Installation of screening on cooling tower sumps will reduce the quantity of large particles that are introduced into the condenser water system. If screens are installed, it will be necessary to inspect them on a regular basis to prevent accumulated debris from restricting or blocking the flow of water from the tower to the chiller.

A comprehensive water treatment program is the most effective means of controlling all forms of chiller system fouling. Water treatment helps to control not only system fouling, but also other system problems, including scale formation, corrosion, and foaming.

To be effective, the program must be specifically tailored to the conditions found in the system to be treated. Water samples must be taken and analyzed to determine the specific problems that exist in the water system and their magnitude. Based on the findings of the analysis, a program is designed that recommends treatments, quantities, and feeding methods. Once started, the program must include periodic testing to determine the effectiveness of the treatment program, and to recommend modifications necessary to provide proper protection to the water system. More frequent testing may be required for facilities whose water supply condition varies frequently.

Most organizations elect to contract all or part of the program to outside firms specializing in water treatment. These firms perform the survey of existing system and water conditions, recommend specific water treatments based on the conditions found, provide guidelines for operators in the use of water treatment chemicals, train in-house personnel in water testing procedures, review the periodic water tests, and recommend changes in the chemical treatment program as required.

Even with a good water treatment program, it will be necessary to bleed off a portion of the circulating condenser water, replacing it with fresh water. Due to the action of cooling towers, a portion of the water circulated in the system is constantly lost to the atmosphere through evaporation and entrainment in the air flowing through the tower. A tower that is well maintained and in good operating condition will typically loose 0.2 percent of the water flowing through it.

The majority of the contaminants normally found suspended in the water are left behind as this water is lost. As a result, the concentration of solids in the water circulated in the system increases. While chemical water treatment will help to keep these solids in suspension, eventually the concentration of solids will become too high and they will precipitate out of the water and form scale on heat transfer surfaces. For this reason, a portion of the circulating water is constantly bled off the condenser water system while the system is operating.

The rate of bleed-off depends on the quality of the water in the system and the rate at which solids are introduced into the system. Too low a rate and the solids will accumulate in the system. Too high a rate and the costs for water and chemical treatment increase. As part of their analysis, the chemical water treatment company can identify the ideal rate of bleed-off.

Monitoring the quality of the water circulating in the condenser and chilled water systems and reviewing the chiller logs will provide an indication of how effectively fouling is being controlled. Additionally, the chiller should be opened and inspected annually or bi-annually, depending on the application, the size of the chiller, and the chiller's history. The inspection will determine if changes are needed in the water treatment program, the rate at which water is being bled off from the system, and what changes in chiller operation need to be made.

How Corrosion Works in Chiller Systems

Corrosion is an electrochemical process that deteriorates metals in chiller systems. Nearly all types of metals used in the systems are impacted by corrosion, including steel, copper alloys, aluminum, and zinc. Corrosion reduces the efficiency of heat transfer surfaces and, if unchecked, leads to the destruction of the equipment. The rate of corrosion within the system is determined by the presence of oxygen in the water, the degree of alkalinity or acidity of the water, the temperature and velocity of the water, the concentration of dissolved and suspended solids, and the quantity of microbiological organisms present in the system.

There are three ways that corrosion attacks the metals found in chiller system: general, localized, and galvanic. Generalized corrosion is ongoing on all metal surfaces exposed to the circulating water in the system. It occurs at a relatively slow rate and is fairly evenly distributed throughout the system. Generalized corrosion is a major problem for system efficiency in that it generates a large volume of oxides that attach themselves to heat transfer surfaces, reducing the efficiency of the system.

While generalized corrosion is primarily an efficiency issue, localized corrosion is more an issue of the equipment's useful life. Localized corrosion occurs when isolated areas of the metals used in the systems become corroded. Other metal surfaces usually are protected so all of the corrosive action becomes concentrated in small areas. This concentrated action results in rapid pitting and perforation of the metal. Localized corrosion is the most serious form of corrosion and is more difficult and costly to correct and repair than others. Considering that most chiller tubes are three thirty-seconds of an inch thick or less, localized corrosion can rapidly eat through and destroy them, allowing refrigerant and water to mix in the system.

The third form of corrosion found in chiller systems is galvanic. Galvanic corrosion occurs when two different metals are in direct physical contact with each other. The dissimilar metals, in the presence of the circulating water, act like a battery and set up a current between them that rapidly corrodes one of the surfaces. The rate of corrosion depends on the level of contamination in the circulating water.

How to Reduce Corrosion

Controlling corrosion requires knowledge of the metals used in a chiller system and an understanding of their susceptibility to corrosion under the operating conditions found in the system. The best defense is a comprehensive water treatment program that specifically addresses chiller system corrosion by reducing the reactiveness of the contaminants in the water and by treating heat metal surfaces by applying a protective film. Adding the proper mix of chemicals in the right quantity will keep internal corrosion to a minimum. As with all treatment programs, water samples will have to be taken periodically to provide feedback of performance. Additionally, the chiller tubes should be inspected either annually or bi-annually for corrosion. While galvanic corrosion can be accelerated through improper water treatment and chiller maintenance, its rate is determine primarily by the design of the chiller.

How Scale Works in Chiller Systems

Scale is a particular type of fouling occurring in chillers. It is highly detrimental to operating efficiency, and occurs when soluable salts found in the water precipitate out and become attached to heat transfer surfaces, particularly on the condenser water side of a chiller. As tower water passes through the chiller, it is warmed, resulting in a decrease in the ability of the water to keep the dissolved minerals in suspension, including calcium sulfate, calcium phosphate, and calcium carbonate. These salts then form a hard layer that adheres firmly to heat transfer surfaces. Essentially a specialized form of fouling, scale is particularly difficult to remove, requiring either chemical or mechanical cleaning of heat transfer surfaces. Scale decreases the efficiency of a chiller by restricting heat transfer between the refrigerant and the condenser water, and increases maintenance costs by requiring periodic shutdown and cleaning of the chiller.

How to Reduce Scale

Three factors contribute to the formation of scale: the temperature rise in the chiller's condenser, the alkalinity or acidity of the water, and the level of calcium hardness of the water. All three can be controlled through the use of good operating and maintenance practices. The temperature rise in the chiller can be limited to the manufacturer's recommended level by maintaining the proper condenser water flow rate. The alkalinity or acidity of the water can be properly controlled through a good water treatment program. The level of calcium hardness of the water can be controlled through water treatment and by properly bleeding off water from the system to reduce the concentration of salts caused by tower water losses.

HOW TO SET UP CHILLER MAINTENANCE TASKS

In addition to the factors of fouling, scale, and corrosion that must be controlled, there is a wide range of elements in chiller operation that require ongoing and periodic maintenance if the unit is to perform efficiently and effectively. The best source for information on what specific tasks need to be performed and their frequency is the chiller manufacturer. The manufacturer is most familiar with the equipment and the installation. What is presented in this section is a general guideline for setting up a maintenance program for chillers. To be most effective, it will have to be modified to match the specifics of the application.

Chiller maintenance activities are typically performed on a daily, weekly, monthly, or annual basis. Figure 6.3 lists typical daily and weekly maintenance activities for vapor compression chillers and their recommended frequencies. Figure 6.4 lists annual maintenance activities. Depending on the critical nature of the operation supported by the chiller, it may be necessary to perform those activities more frequently.

Chiller Leaks

One of the most critical aspects of chiller maintenance is to ensure the integrity of the chiller system. In high-pressure systems, refrigerant will leak to the atmosphere. In low-pressure systems, air and moisture can be drawn into the chiller refrigerant system.

Figure 6.3
Daily and Weekly Maintenance Activities for Vapor Compression Chillers

1. Building/facility: _____

2. Room #: _____   3. Chiller #: _____

4. Dates: _____

Activity	Day of Week						
	1	2	3	4	5	6	7
Complete chiller maintenance log							
Check general operating conditions							
Check for unusual noises and vibration							
Inspect for oil leaks							
Inspect for water leaks							
Check oil level							
Check refrigerant level							
Inspect purge unit operation							
Log purge unit cycle counter reading							
Sample condenser water							
Inspect water treatment system							
Inspect valves for leaks							
Check filters and strainers							
Check pH of chilled water							
Check pH of condenser water							
Inspect V-belts for wear and fit							
Inspect drive couplings for wear							
Inspect sheaves for wear							

Figure 6.4
Annual Maintenance Activities
for Vapor Compression Chillers

1. Building/facility: _____

2. Room #: _____ 3. Chiller #: _____

4. Date: _____

	Completed	Comments
Sample oil for testing	☐	_____
Change compressor oil and filter	☐	_____
Sample refrigerant for testing	☐	_____
Change refrigerant filter	☐	_____
Test operation of purge unit	☐	_____
Calibrate controls	☐	_____
Test linkages for tightness	☐	_____
Test safeties		
Chilled water low temperature	☐	_____
Chilled water flow	☐	_____
Condenser water flow	☐	_____
Refrigerant low pressure	☐	_____
Refrigerant high pressure	☐	_____
Low oil pressure	☐	_____
Meg motor windings	☐	_____
Inspect all wiring, starters, & disconnects	☐	_____
Test all indicator lights	☐	_____
Check motor to compressor alignment	☐	_____
Inspect motor mounts and bolts	☐	_____
Inspect chiller tubes for scale and fouling	☐	_____
Inspect case for rust	☐	_____
Eddy current test tubes	☐	_____

In both systems, leaks between the refrigerant and the condenser or chilled water systems can result in water entering the refrigerant system. Finding and correcting leaks will help the facility to comply with clean air laws, reduce the use of refrigerants, improve the operating efficiency of the chillers, and reduce wear on chiller components.

For low-pressure chillers, the greatest concern is with air and moisture leaking into the refrigerant system. Air is a noncondensable gas and greatly reduces the chiller's capacity and efficiency. Moisture interferes with the operation of the refrigerant system and can form acids that can quickly damage bearings, gears, and the windings on motors in hermetically sealed chillers.

Low-pressure chillers should be constantly monitored for leaks. By tracking system operating pressures and ambient conditions, operators can identify changes in the operation of the system resulting from air in the refrigerant. Additionally, refrigerant samples should be drawn at least once per year and sent to a lab for analysis.

All low-pressure chillers should be equipped with high-efficiency purge units. The purpose of the purge unit is to remove noncondensable gases from the refrigerant. While the operation of all purge units results in the loss of some refrigerant, high-efficiency units greatly reduce the losses. To detect leaks, purge unit operation should be monitored through a run-time monitor or by a simple frequency counter. Monitoring purge unit operation will allow operators to detect leaks as they develop.

There is an additional leak concern for high-pressure chillers—loss of refrigerant. This reduces the operating efficiency of the chiller, increasing energy requirements. Replacement refrigerants are a significant expense, particularly for chillers that use CFC-based refrigerants.

Tube Testing

One of the most common areas in which chillers develop leaks, and one of the most expensive items to repair, are the chiller's heat exchanger tubes. Leaks in the tubes develop slowly, primarily as a result of corrosion, erosion, pitting, and thermal stresses. Detected early, the defective tube can be plugged or replaced before it starts leaking.

At least once every two years, chillers should be taken off line and opened for an inspection of the heat exchanger tubes. Chillers serving critical applications, large chillers, and older chillers need to be tested every year. While visual inspection can identify problems such as heavy fouling and corrosion, it cannot detect most problems that lead to leaks. Detection of problems as they are developing requires the use of an eddy current tester.

Eddy current testers use an alternating current to produce a magnetic field around a probe that is inserted into the heat exchanger tubes. As the probe travels through the tube, it induces an eddy current in the tube. Defects in the tube, such as corrosion, pitting, and thinning, cause variations in the eddy current. An experienced operator, by monitoring the changes in current, can determine the location and magnitude of the defect in the tube. If the detected defects are severe enough, repairs can be made while the chiller is shut down.

Results of eddy current testing should be compared to past testing. By maintaining the records of the testing results, the condition of the heat exchanger tubes can be tracked over time, thus giving you an indication of how rapidly the tubes are deteriorating and how soon repairs will be required.

Refrigerant Testing

The condition of the refrigerant in the system is critical to the proper operation and life of the system. Even with normal use, refrigerant can become contaminated with moisture, rust particles, and other contaminants. These contaminants can lead to the buildup of sludge at the bottom of the chiller's evaporator shell, the formation of acids that attack metal surfaces within the chiller, and the degradation of heat transfer surfaces.

Rule of Thumb: At least once a year, a refrigerant sample should be drawn and forwarded to a qualified lab for testing. The refrigerant sample gives a very good view of the condition of the chiller system internals, detecting potentially damaging conditions before they result in the need for a major system overhaul.

Oil Testing

Oil testing, like refrigerant testing, provides information on the condition of the chiller system's internals. *Rule of Thumb:* At least once a year, an oil sample should be drawn and sent to a qualified lab for analysis. The testing results will show the level of contaminants in the compressor oil, including metallic particles that would indicate wear in components such as bearings and gears. Oil sample results should be logged and tracked over time. Rising levels of particular contaminants are early warning signs of developing wear problems within the chiller.

ABSORPTION CHILLER MAINTENANCE CONSIDERATIONS ─────────

Absorption chillers require a number of maintenance activities, many similar to the ones performed on vapor compression chillers. Figure 6.5 lists typical daily and weekly maintenance activities for absorption chillers and their recommended frequencies. Figure 6.6 lists recommended annual maintenance activities. As with maintenance activities for vapor compression chillers, these activities represent the minimum level of maintenance to be performed on these chillers. Additional activities and shorter maintenance intervals may be required as a result of the particular installation.

While maintenance is important for the proper operation of vapor compression chillers, it is critical to the operation of absorption chillers. Absorption chillers have gained a reputation as being unreliable and prone to frequent breakdowns. More often than not, problems with the operation of the chillers can be traced to a lack of proper maintenance. Two particular maintenance concerns are maintaining the proper vacuum within the shellside of the absorber and controlling corrosion within the chiller.

Figure 6.5
Daily and Weekly Maintenance Activities
for Absorption Chillers

1. Building/facility: _____

2. Room #: _____   3. Chiller #: _____

4. Dates: _____

Activity	Day of Week						
	1	**2**	**3**	**4**	**5**	**6**	**7**
Complete chiller maintenance log							
Check general operating conditions							
Check for unusual noises and vibration							
Inspect for water leaks							
Inspect solution pump motor							
Inspect purge unit operation							
Log purge unit cycle counter reading							
Inspect purge unit belt							
Sample condenser water							
Inspect water treatment system							
Inspect valves for leaks							
Check filters and strainers							
Check pH of chilled water							
Check pH of condenser water							
Sample lithium bromide for testing							

Figure 6.6
Annual Maintenance Activities
for Absorption Chillers

1. Building/facility: _____

2. Room #: _____   3. Chiller #: _____

4. Date: _____

	Completed	Comments
Test operation of purge unit	☐	_____
Change purge unit pump oil	☐	_____
Calibrate controls	☐	_____
Test safeties		
Chilled water low temperature	☐	_____
Chilled water flow	☐	_____
Condenser water low temperature	☐	_____
Condenser water flow	☐	_____
Inspect all wiring, starters, & disconnects	☐	_____
Test all indicator lights	☐	_____
Inspect chiller tubes for scale and fouling	☐	_____
Inspect case for rust	☐	_____
Eddy current test tubes	☐	_____

Maintaining the Vacuum

There are two actions that contribute to the loss of vacuum in an absorption chiller—leaks and the formation of hydrogen. A vacuum is required to permit the refrigerant, typically water, to vaporize at low saturation temperatures. Any loss of vacuum and the capacity and efficiency of the absorber are reduced. With the absorber under vacuum, any leaks in the system result in air entering the system. In addition to reducing capacity and efficiency, air leaks introduce oxygen into the system. Lithium bromide, the absorbent used in most systems, is a salt and becomes extremely corrosive in the presence of oxygen. Uncorrected vacuum leaks will quickly lead to corrosive damage to the steel of the chiller.

Hydrogen gas is generated naturally as a result of the reaction of the lithium bromide with the steel surfaces of the chiller. Hydrogen is a noncondensable gas that, unless removed, will reduce the capacity and efficiency of the system.

The level of noncondensable gases in the absorption system is controlled most effectively through the use of a purge unit. The purge unit, similar to the ones used in vapor compression chillers, collects and removes noncondensables from the absorption system, thus helping to maintain chiller capacity and reduce the rate of internal corrosion.

The purge unit also helps to reduce the risk of crystallization. Crystallization occurs when the lithium bromide begins changing from a liquid to a solid, forming a crystalline slurry mixture that plugs chiller pipes, orifices, pumps, and other areas. While crystallization does not usually harm the chiller, it does make it temporarily inoperable. It is corrected by applying heat to the areas that are plugged so that the crystallized lithium bromide can be changed back to a liquid.

Purge unit operation should be constantly monitored. Increasing run times are an indication that the system is developing a leak and will require testing and corrective action during the next annual inspection.

Controlling Corrosion

In addition to removing air from the refrigerant, it is necessary to maintain the chemical properties of the lithium bromide within recommended limits to control internal corrosion. There are two maintenance activities involving the lithium bromide that are required in order to control internal corrosion: addition of a corrosion inhibitor and regular testing.

The use of a corrosion inhibitor is important to limit the effects of the lithium bromide on the internal steel of the chiller. Properly applied in conjunction with an efficient purge unit, it can limit the rate of internal corrosion to an acceptable level.

Regular testing of the lithium bromide is required to maintain the proper alkalinity of the absorbent and avoid damage and performance loss in the chiller system. *Rule of Thumb:* A sample of the absorbent should be drawn at least one every month and sent to a qualified lab for analysis. As with other samples, the data from the analysis should be tracked over time to watch for slowly developing trends.

CHILLER OPERATIONS FOR ENERGY MANAGEMENT ————————

In addition to maintenance activities that can be implemented to improve chiller operating efficiency, operational practices can be put in place to further reduce chiller energy use. The operational practices discussed here can be implemented with minimal first costs, provided that the chiller system has the proper controls, or they may require a minor investment in control system modifications. Even if modifications are required, the cost of those modifications is so low in comparison to the savings produced that the payback period for those chiller modifications is measured in months.

Chiller Load Allocation

This operational practice applies to facilities having a central chiller plant consisting of at least two chillers connected in parallel. Normal practice in these facilities is to have several chillers on line at the same time, with the cooling load split evenly between them. Even when the load is sufficiently low that one chiller could carry it, operators keep two running in order to maintain reserve system capacity and so that there would be no interruption of service should one of the chillers shut down.

Unfortunately, the practice of keeping two chillers operating at part load when one chiller could carry the full load wastes energy. Chiller efficiency falls off with decreasing load. For example, a centrifugal chiller operating at 50 percent full-load capacity typically uses 60 to 65 percent of the full-load energy. The practice of operating two chillers also increases wear and maintenance requirements as chiller run time increases. While it does help protect against the failure of a single chiller, the benefit is more than offset by these increased costs in all but the most critical applications.

In its simplest form, chiller load allocation is the practice of starting and stopping chillers to more closely match on-line chiller capacity with cooling loads. Operators monitor the loads on the chiller system and start or stop additional chillers as required. The base load for the facility is carried by the most efficient chiller. When the load is increasing, additional chillers are brought on line, starting with the highest-efficiency idle chiller. When the loads are decreasing, chillers are taken off line starting with the lowest efficiency chiller.

In more complex applications, loads are optimized across multiple chillers operating at the same time. When loads are low, only the most efficient chiller is operated. As the load increases, the next most efficient chiller is brought on line. The most efficient chiller continues to operate at full capacity while each increment of increasing load is added to the next most efficient chiller. If the load decreases, the increment is subtracted from the lower-efficiency chiller.

Implementing chiller load allocation is primarily procedural. Chiller control systems must be set up to support the load allocations, but little or no additional hardware is required.

Condenser Water Temperature Reset

For years, the standard operating procedure for chiller condenser water temperature has been to maintain a constant cooling tower water return temperature, typically 85 degrees, regardless of load on the chiller or outside temperatures. To maintain cooling tower return water at this temperature, tower fans are cycled and tower bypass valves are opened. This practice is primarily the result of a belief that the chiller operates best at a constant condenser water temperature and that energy will be saved by turning cooling tower fans off.

In practice, energy will be conserved by reducing the tower water return temperature. *Rule of Thumb:* For every ten-degree reduction in condenser water supplied to the chiller, the chiller's compressor energy requirements will be reduced by approximately 15 percent. With chillers and cooling towers sized for the peak load, systems typically spend more than 95 percent of the time operating at reduced loads and conditions when condenser water temperatures can be reduced.

All chillers can be operated with reduced condenser water temperatures. How low the temperatures can be reduced depends on the model of the chiller; manufacturers list the recommended operating ranges for their units. Thermal stresses within the chiller's heat exchanger will be increased at these lower temperatures, but as long as temperatures are kept within the recommended range of the manufacturer, these stresses will not damage the chiller. Similarly, there is a slight increase in the rate of scaling in the chiller's heat exchanger that can be controlled through proper water treatment.

Absorption chillers can also be operated at lower condenser water temperatures. While these chillers are sensitive to condenser water temperatures and salt up if they are not properly controlled, salting is not a problem as long as they are operated within the manufacturer's recommended temperature range.

Condenser water temperature reset will require some modifications to the chiller's and cooling tower's control system. The modifications are minor and their cost will typically be recovered in less than one cooling season.

Chilled Water Reset

Like condenser water, the standard operating procedure for chiller cooling water temperature has been to maintain a constant supply water temperature, typically 42 to 44 degrees, regardless of the load on the chiller. In practice, the supply water temperature required to adequately cool and dehumidify a facility depends on capacity of the system and the cooling load. When the system is operating at less than peak loads, it is possible to reduce chilled water supply temperatures and still meet these load requirements. With a reduction of approximately 1.5 to 2.0 percent in compressor energy requirements for each degree increase in chilled water supply temperature, and with the system operating below its maximum design load most of the time, it is possible to achieve major energy savings just by increasing the chilled water supply temperature.

Changing the chilled water temperature requires no modifications to the chiller.

Operators can set the supply water temperature each day based on the projected load on the system. However, raising chilled water supply temperatures does decrease the control the system has over space humidity conditions. To achieve the maximum savings while maintaining adequate control over space temperatures and humidity levels, space conditions will have to be monitored. Feedback from those conditions then can be used to automatically reset chilled water supply temperatures. ***Rule of Thumb:*** Typical payback for such a system is less than one season.

Indirect Free Cooling

Under certain conditions it is possible to obtain a limited amount of "free" cooling from a chiller system. Under this mode of operation, the cooling tower and chilled water systems are operated normally, but the chiller is not turned on. Instead, refrigerant migration valves in the chiller are opened to equalize pressure in the chiller's evaporator and the condenser. Since the condenser is at a lower temperature than the evaporator, freon is condensed and migrates to the evaporator where it vaporizes. As a result, heat is transferred from the chilled water system to the condenser water system where it is dissipated by the cooling tower. Under this mode of operation, the cooling produced by the system is typically 10 percent of the full-load rating of the chiller.

An indirect free cooling system produces chilled water that is approximately ten degrees warmer than the water supplied to the chiller from the cooling tower. The systems are best suited for facilities that have winter cooling loads that are less than 10 percent of the chiller's peak capacity and are located in climates where it is possible to obtain 40 to 45 degree cooling tower water temperatures. Only minor modifications are required to set up indirect free cooling.

COOLING TOWER MAINTENANCE CONSIDERATIONS

Cooling towers are one of the most overlooked items in building chiller systems. The problem is that most maintenance problems associated with cooling towers are not readily apparent. These maintenance problems generally result in a slow but steady decline in tower performance. While the tower may continue to operate, it does so at reduced capacity and efficiency. Unless operators are paying close attention to chiller system operation, these declines continue undetected until the tower is called upon to operate on a peak cooling day and cannot meet the load. Even then, it is generally the chiller system that is blamed for the lack of capacity, not the cooling tower.

Cooling towers play a critical role in the overall efficiency of a chiller system. Unfortunately they are high-maintenance items as a result of both the operation that they are required to perform and the environment in which they must operate. A cooling tower can be thought of as a highly efficient air washer. As a result of the water flow through the open tower, dust, pollen, debris, organisms, and chemicals are scrubbed from the air and introduced into the condenser water system. Left unchecked, these contaminants result in increased chiller head pressures, increased compressor energy use, and fouled heat exchanger surfaces. Some contaminants, such as Legionella, can create health hazards.

How important is the condition of the cooling tower to the efficient operation of the chiller system? Consider what happens to a 1,000-ton chiller system that, because of inefficient tower operation, returns condenser water to the chiller at one degree above normal. The resulting annual energy penalty amounts to more than $50,000 at an average rate of $.06 per kWh. And that was just for a one degree increase. Most systems that are not properly maintained have higher condenser water return temperatures, resulting in even greater energy penalties.

Tower Maintenance Requirements

Water treatment is the first line of defense against the elements that are at work to reduce tower efficiency. Good water treatment reduces the impact of fouling and scale, both of which can rapidly decrease the performance and efficiency of a tower.

There are two major elements required for efficient cooling tower operation—uniform water distribution and uniform air flow. Anything that interferes with either of these elements reduces the performance of the tower. Therefore maintenance should focus on the tower's components that enhance these elements—the fill, distribution system, drift eliminators, and fans. In towers that have been neglected, it is not uncommon to find broken fill, damaged or missing drift eliminators, clogged pipes and nozzles in the distribution system, and improperly operating fans; all problems that contribute to increased chiller system energy use, and all problems that could have been detected and corrected through a comprehensive inspection and maintenance program.

One additional area that requires ongoing inspection and maintenance is the tower structure. Although the structure is less likely to impact system efficiency than other tower components such as the tower fill, it is essential that tower structure be maintained in order to prevent premature failure.

Fill Material

The tower fill is the most critical component in the operation of the cooling tower. Its function is to expose as much surface of the recirculating water to the passing air so that heat transfer can be enhanced. Various materials have been used in tower construction over the years, including wood, cement, ceramic, asbestos, metal, and plastic. Most new tower designs make use of plastic fill material as it resists scaling, is light-weight, and low in cost.

In older tower designs, the fill consisted of a series of splash bars or grids that caused the water to fall turbulently, continually exposing new surfaces to the air flow. Newer tower designs cause the water to flow over sheets of fill material in thin films for maximum air flow exposure. In both designs, if the tower's fill material is damaged, out of place, missing, or badly scaled, the surface area of the water that will be exposed to the air flow through the tower will be reduced, cutting tower performance.

Rule of Thumb: Periodic inspections while the tower is operating, with careful observation of the water flow through the fill material, will help to identify damaged areas in the fill. A detailed inspection at the end of the cooling season with the tower shut down will further identify areas where the fill material needs replacement. Another detailed inspection with the tower shut down should be conducted in the spring before the cooling season starts to identify damage from ice.

Drift Eliminators

As air flows through the tower, particularly when the tower fans are operating on high speed, some of the recirculating water becomes entrained as droplets in the air flow and is carried off. Called drift, this water loss contributes nothing to the efficiency of the tower while increasing makeup water and water treatment chemical requirements. To reduce the amount of drift released from a tower, manufacturers place an assembly, called a drift eliminator, in the airstream to remove the entrained water droplets. In a well-maintained tower, drift can be reduced to less than 0.2 percent of the recirculating water flow rate.

With time, drift eliminators can become blocked by scale and biological buildup. Panels in the eliminators can sag or deteriorate. The result is partially blocked air flows and excessive water loss from the tower, which can be seen in the form of wet or stained surrounding surfaces. While improperly functioning drift eliminators do not usually increase energy use, they do result in an increase in the tower's water use and an increase in the chemicals required to treat tower water.

Inspections conducted both while the tower is operating and while it is shut down will help to identify areas where drift eliminators are in need of repair or replacement. The drift eliminators must completely enclose the plenum area. Look at the surrounding area for stains from water and mineral deposits thrown out of the tower. Watch the tower discharge area for falling water droplets.

Water Distribution System

The operation and condition of the tower's distribution system also has a major impact on the tower's performance. Ideally, the water entering the tower should be evenly distributed as it enters the fill. This requires proper balancing of the water distribution lines and proper operation of the distribution nozzles. Any uneven flow will result in reduced tower capacity and efficiency.

One of the most common problems found in the distribution system is improperly operating nozzles. Mineral deposits and biological fouling can partially or fully block the flow through a nozzle. It is not uncommon to find that one or more nozzles have broken off.

Another common problem is uneven water flow through tower cells as the result of improper balancing or clogged distribution lines. Most systems use balancing valves to regulate the flow through the distribution system, resulting in equal water flow into all tower cells. Frequently the valves are improperly set, resulting in unbalanced flow. It is not uncommon to find all balancing valves partially closed, unnecessarily throttling the water flow and increasing pumping energy costs.

While the tower is operating, observe the spray patterns produced by all nozzles. Patterns should be symmetrical and equal between nozzles. Check the level of the water in the sump of each of the tower's cells. Unequal water levels may be an indication that the flow rates through the distribution system are improperly balanced.

Fans

Cooling tower fans are expected to operate in the worst possible conditions for mechanical equipment: high temperatures, and high humidities while being exposed to the elements. The result is that for proper operation, tower fans need ongoing maintenance. Fan operation is critical to the use of the tower as most towers will operate at only 10 percent of their capacity in the event of a failure of the fan system.

The most common problem with tower fan operation is the drive mechanism. Most tower fans operate at relatively low speeds, requiring the use of reduction gears and drive belts. These drive components are subject to wear and failure. About the only way to avoid having to shut down the chiller system due to a drive system component failure is to conduct frequent inspections of the components, both while the tower is operating and while it is shut down.

Tower Structures

The operation and maintenance instructions for cooling towers are supplied by the manufacturer when a tower is new. The maintenance instructions presented in this section are intended to supplement those supplied by the manufacturer. In all cases it will be necessary to modify both sets of instructions to match the specifics of the tower installation. For example, if the tower is located in a particularly dusty or dirty environment, it would be advisable to conduct more frequent inspections for contaminants and silting.

Tower maintenance starts with documentation. All inspections, reading, maintenance actions, and breakdowns must be documented in a formal log. The log not only confirms what activities have been performed and when they were performed, but also provides a history of tower operation that can be used to diagnose and project long-term trends in tower performance. To ensure that inspections and routine maintenance are thorough, checklists should be used and filed in the tower log.

Cooling tower structures are expected to operate under the worst possible environmental conditions. In addition to being exposed to the elements, tower structures are subject to constant wetting from the normal operation of the tower. Water, chemicals, and biological organisms attack metal and wood components, resulting in corrosion and decay. Without proper maintenance attention, tower structures can fail or become structurally unsafe well before the end of their normal expected life.

Rule of Thumb: Tower structures should be closely inspected before the start and after the end of the cooling season. During the cooling season, tower structures should be inspected at least every two weeks. Particular attention should be paid to areas where dissimilar materials, such as two different types of metal, or metal and wood, are in contact with one another.

Some tower structural problems have a direct impact on the operating efficiency of the tower. Damaged fill support structures, loose fitting access doors, damaged casings, missing deck boards—all contribute to tower losses. Other tower structural problems may not impact the efficiency of the unit but do have a direct impact on the safe operation and life of the tower.

Tower Inspection

Tower inspection programs are all similar in scope regardless of the size and type of tower. While the tower is operating, it should be inspected at least every two weeks, and more frequently in critical applications. Not all components of the tower can be inspected with the tower in operation, so additional inspections will have to be completed during the off season or during scheduled tower shutdowns. *Rule of Thumb:* At least two inspections should be completed each year while the tower is shut down—one in the fall after the end of the cooling season, and one in the spring before the cooling season has begun. It is best to schedule the spring inspection at least one month before the tower is to be placed into operation in order to allow time to correct maintenance problems found during the inspection.

Figure 6.7 lists weekly and monthly maintenance activities for cooling towers. Figure 6.8 lists annual maintenance activities. All completed inspections should be filed so that trends in tower condition can be identified over time. The items listed in the figures are considered to be the minimum requirements for a thorough tower inspection program and should be modified to match the requirements of the particular application.

The most commonly used pump type for circulation of both the condenser and chilled water in the chiller system are centrifugal pumps. Centrifugal pumps are known for their long life and low maintenance requirements. Life expectancies of fifteen years can easily be achieved by following the manufacturer's recommendations for maintenance. Although performing routine maintenance will not necessarily reduce the energy requirements for pumping, routine maintenance will reduce the chances for shutdown of the entire chiller system as the result of a pump failure.

Most maintenance activities for centrifugal pumps focus on three aspects of the pump installation: bearings, seals, and alignment. The most important factor in bearing life is applying the right type and quantity of lubricant. Oil bearings must be inspected frequently to ensure that their oil level is kept within the recommended range. Grease lubricated bearings require less frequent inspections. The most common maintenance mistake made with grease lubricated bearings is overlubrication. Too much grease causes the ball bearings to skid or slide, damaging the bearings' race. Follow the manufacturer's lubrication recommendations to prevent both over and underlubrication.

Figure 6.7
Weekly and Monthly Maintence Activities for Cooling Towers

1. Building/facility: _____

2. Location: _____   3. Tower #: _____

4. Dates: _____

Activity	Week of Month			
	1	2	3	4
Check general operating conditions				
Check for unusual noises and vibration				
Check tower basin water level				
Check rate of water bleed-off				
Inspect basin for debris				
Observe water flow over fill				
Observe spray pattern from nozzles				
Inspect fan system for vibrations				
Inspect structure for corrosion				
Clean air intake screens				
Inspect drift eliminators for damage				
Inspect area for excessive drift				
Flush debris from sump				
Clean sump strainer				
Inspect sump for leaks				
Inspect fan drive belts				
Test operation of float valve				
Inspect fan deck				

Figure 6.8
Annual Maintenance Activities for Cooling Towers

1. Building/facility: _____

2. Location: _____   3. Tower #: _____

4. Date: _____

	Completed	Comments
Inspect fill for damaged or missing sections	☐	_____
Inspect fill for fouling and scale	☐	_____
Inspect drift eliminators for damage, misalignment	☐	_____
Inspect drift eliminators for fouling and scale	☐	_____
Inspect basin for corrosion	☐	_____
Inspect tower finishes for corrosion	☐	_____
Inspect structural members for corrosion	☐	_____
Inspect safety railings, stairways, and walkways	☐	_____
Inspect distribution system piping	☐	_____
Clean distribution nozzles	☐	_____
Check distribution system balancing	☐	_____
Inspect fan blades for damage	☐	_____
Inspect drive belts and tighten	☐	_____
Change oil in gearboxes	☐	_____
Inspect wiring and fan controls	☐	_____

Pump seals come in two major types—packing glands and mechanical seals. Packing glands require periodic inspections for leaking. If the packing gland is found to be leaking, it should be tightened carefully to prevent damage to the seal and the pump shaft from overtightening. Mechanical seals require periodic inspection for leaking. Most cannot be repaired.

Pump seal failure is typically the result of either contaminants in the fluid being pumped or misalignment between the pump and its drive motor. Contaminants in the fluid can get trapped between the seal and the rotating pump shaft, resulting in excessive wear or scoring. Poorly aligned pumps and motors put excessive side-play on seals, resulting in early failure.

Pump alignment is also important to reduce damage from vibrations, wear in couplings, and wear of pump and motor bearings. The better the alignment, the less wear and stress in the system.

Figure 6.9 lists recommended weekly and monthly maintenance activities for centrifugal pumps. These activities were developed for use with pumps of five horsepower or larger, although the reliability of any centrifugal pump will be enhanced by implementing these activities.

In addition to the weekly and monthly activities, centrifugal pumps should be inspected once a year. During that inspection, additional maintenance activities are performed. Figure 6.10 lists centrifugal pump annual inspection and maintenance requirements.

IMPLEMENTING THE ENERGY MAINTENANCE PROGRAM

One of the most difficult problems to overcome when establishing a maintenance program for chiller systems is knowing where to start. If maintenance has been substandard or neglected for a period of time, all chiller systems within the facility will be in need of major maintenance attention. All could benefit from the program. And the maintenance program would reduce the energy requirements of all of the systems. The question is, where do you start when you need to do everything?

Given the limited resources that most organizations have, it will be impossible to implement a comprehensive maintenance program for all chiller systems throughout the facility at the same time. If reliability is the major priority, start with the systems that serve the most critical applications. If energy conservation is the major priority, start with the largest systems, or the systems that have the highest number of operating hours per year. If improved performance is the major priority, start with the systems that are in the worst condition.

Lay out a phased plan for implementing the program that is within the manpower and budget of the maintenance operation. While it may take several years for the program to become fully implemented, the benefits of reduced energy use and increased system reliability will be noticed during the first year. As more chiller systems come under the operation of the program, the number of breakdown maintenance calls will decrease, allowing even more time to be dedicated to the program.

Figure 6.9
Weekly and Monthly Maintenance Activities for Centrifugal Pumps

1. Building/facility: _____

2. Room #: _____   3. Pump #: _____

4. Type of system: _____

5. Dates: _____

Activity	Week of Month			
	1	2	3	4
Check general operating conditions				
Check for unusual noises and vibration				
Inspect packing glands & seals for leaks				
Check bearing oil level				
Inspect for leaks				
Check bearing temperature				
Check bearing lubricant				
Clean strainers				

Figure 6.10
Annual Maintenance Activities for Centrifugal Pumps

1. Building/facility: _____

2. Room #: _____   3. Pump #: _____

4. Type of system: _____

5. Date: _____

	Completed	Comments
Perform vibration analysis	☐	_____
Inspect pump housing	☐	_____
Remove, clean, and repack bearings	☐	_____
Remove packing and inspect shaft seal	☐	_____
Re-align pump and motor	☐	_____
Inspect coupling for wear	☐	_____
Retorque mounting bolts	☐	_____
Inspect motor wiring, starter, and disconnect	☐	_____
Meg motor windings	☐	_____

Start the maintenance program for each chiller system with a very detailed inspection, examining the systems and components in even more detail than normal. If maintenance has been neglected, it will be necessary to identify all of the existing deficiencies in the systems so that a plan can be laid out for their correction. Remember to keep accurate records of what was found during the inspection, and thorough logs of corrective actions taken.

SUMMARY

A well-planned and implemented maintenance program for chiller systems can reduce the annual energy and breakdown maintenance costs for the system by as much as 50 percent. Additional savings will be realized from reduced system downtime. In nearly all applications, the program startup costs will be recovered during its first season of operation.

Chapter 7

Boiler Systems

This chapter presents information and guidelines for maintaining boiler systems. This information has been generalized for steam systems typically found in building heating and plant processing applications. Managers of facilities that have boiler systems that are used for special purposes or under unusual conditions will need to modify these guidelines to meet the requirements for their specific application. If modifications are required, it is recommended that the manufacturers of the components used in those boiler systems be contacted for additional information. Manufacturers are the ones most familiar with the maintenance requirements for their equipment and have published detailed listings of maintenance requirements and maintenance procedures.

For the purposes of this chapter, boiler systems are defined to include the boilers and related generation equipment, the steam distribution system, and the condensate return system. Other components, such as steam coils and heat exchangers, will be included in subsequent chapters.

BOILER SYSTEM ENERGY MAINTENANCE CONSIDERATIONS

Those who work daily with boiler systems understand that maintenance is the most effective tool for reducing problems and improving performance. In spite of this, few organizations have an effective boiler system maintenance program in place. As a result, there are thousands of boiler failures and interruptions of service each year. It has been estimated that between one-half and two-thirds of all boiler failures and steam service interruptions are the direct result of poor maintenance. The financial drain of equipment replacement and loss of use of the facilities costs organizations more than ten times what they would have spent on a comprehensive boiler system maintenance program.

Service interruptions and premature equipment failure, as costly as they are, are small in comparison to the cost of the energy wasted due to a lack of boiler system maintenance. Studies of boiler systems have demonstrated that improvements in operating efficiencies can be achieved in nearly every system examined. In well-maintained systems, efficiency improvements of 1 to 2 percent were common. In moderately maintained systems, changes in maintenance procedures were identified that would result in improvements in operating efficiency of between 2 and 5 percent. In poorly maintained systems, maintenance programs typically improved efficiencies by 10 to 30 percent.

How significant is this to energy costs? It is very significant when one considers that the typical boiler burns more in fuel in one year than it cost to purchase the boiler. Even a 1 percent improvement in boiler operating efficiency produces large savings over the course of a year. ***Rule of Thumb:*** The savings produced by a 1 percent increase in operating efficiency of a 100,000 lb per hour boiler burning oil at one dollar per gallon amounts to more than $113,000 per year. Figure 7.1 shows the savings produced for each 1 percent improvement in efficiency based on the firing rate of the boiler and the cost of the fuel.

The savings identified in Figure 7.1 can be applied to improvements in operating efficiency made anywhere in the boiler system. Improvements to the steam distribution system and the condensate return system are as important as improvements to the boiler itself. In many facilities, the potential for improvement is even greater in these areas due to a lack of maintenance. The boiler, however, is the most common starting point when establishing a maintenance program to improve the operating efficiency of a complete boiler system.

Even new boilers with sophisticated computer control systems can benefit from a comprehensive maintenance program. Too often, owners and operators assume the automatic features on these boilers will keep them operating at peak efficiencies without additional intervention. Unfortunately, many of the factors that require periodic maintenance are not eliminated by these automatic controls and watchdog devices. Without proper maintenance, these boilers will also rapidly decline in efficiency over time, resulting in increased fuel use and unsafe operations. Simply stated, all boilers need maintenance. Without it, they will operate inefficiently and will eventually fail.

THREE METHODS OF RATING BOILER EFFICIENCY

The primary objective of any boiler maintenance program is to keep the boilers operating safely and efficiently. Implementing operational and maintenance practices to improve boiler operating efficiency does not result in a decrease in boiler safety. Similarly, implementing operational and maintenance activities to enhance safety will not result in decreased boiler efficiency. Good operating and maintenance procedures will enhance both.

Boiler efficiency is influenced by many factors, including the type of fuel used, the load on the boiler, the design of the boiler, the condition of the boiler, and how the boiler is being operated. Average full-load operating efficiencies for boilers typically fall between 75 and 90 percent for units that are in good condition and are being properly operated. Higher efficiencies can be obtained through the use of sophisticated controls

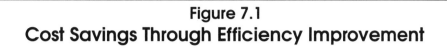

Figure 7.1
Cost Savings Through Efficiency Improvement

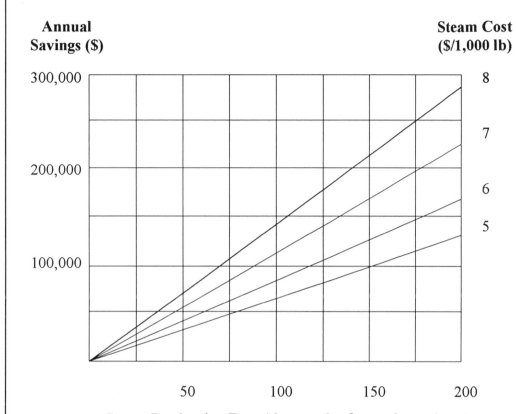

Annual Savings ($)

Steam Cost ($/1,000 lb)

Steam Production Rate (thousands of pounds per hour)

and recovery equipment. Lower efficiencies are generally the warning sign that changes are needed in boiler operating and maintenance practices.

Boiler efficiency is also influenced by how it is defined. There are three major ways to define efficiency: thermal efficiency, combustion efficiency, and fuel-to-steam efficiency. Thermal efficiency rates how well the boiler transfers heat from the combustion gases to the steam, excluding radiation and convection losses. Combustion efficiency rates the effectiveness of the burner to use fuel, excluding radiation and heat transfer losses. Fuel-to-steam efficiency rates the overall effectiveness of the boiler in converting fuel energy into steam energy.

Of the three methods of rating boiler efficiency, the most meaningful one is the fuel-to-steam efficiency rating. Since it is a ratio of the energy output of the boiler divided by the energy input, the fuel-to-steam rating takes into consideration all losses associated with the operation of the boiler, including combustion, stack, heat transfer, radiation, and convection losses. For the facility manager, fuel-to-steam ratings are the only true measure of efficiency as the objective of a boiler maintenance program for energy conservation must seek to reduce all boiler losses.

There are a number of opportunities for improving the fuel-to-steam efficiencies of boilers through good operating and maintenance practices. These include activities that address the boiler itself, the water supplied to the boiler, and the fuel burned in the boiler. Some are ongoing while other require periodic implementation. None require a significant investment in new equipment. If all are properly implemented, you can expect a reduction in boiler energy requirements of at least 5 percent—more than enough to pay for the cost of implementing the maintenance program within the first year.

USING THE BOILER OPERATING LOG

A boiler maintenance program begins with an operating log. The purpose of the log is to assist operators and engineers in recognizing performance deviations that may influence boiler operations, and diagnosing their causes. To be effective, log entries must be made and reviewed on a regular basis. Spotty entries do not provide sufficient information to identify performance trends. For many facilities, daily entries into the log will be sufficient. Entries may need to be made on an hourly basis for large boilers, typically those with steam capacities of 50,000 lbs/hr or more.

Similarly, even the most thorough log must be reviewed on a regular basis if trends are to be identified, particularly before they turn into serious and costly maintenance problems. Monthly review of operating logs is usually sufficient for most boiler installations. For boilers with steam capacities of 50,000 lbs/hr or more, log books should be reviewed at least weekly.

The information that is typically recorded in the boiler operating log includes data on conditions within the boiler, the feedwater system, the combustion air system, the boiler exhaust, and the steam system. Use Figure 7.2 to log boiler operating data. The items listed in the figure are suitable for use with most boilers found in building heating

Figure 7.2
Boiler Log

1. Building/facility: _____

2. Room #: _____   3. Boiler #: _____

Date								
Time								
Steam pressure								
Steam flow rate								
Condensate return temperature								
Feedwater temperature								
Feedwater pressure								
Makeup water meter reading								
Fuel supply temperature								
Fuel supply pressure								
Fuel flow meter reading								
Flue gas temperature								
Flue gas oxygen								
Forced draft fan inlet temperature								
Check water level	☐	☐	☐	☐	☐	☐	☐	☐
Check flame appearance	☐	☐	☐	☐	☐	☐	☐	☐
Check stack appearance	☐	☐	☐	☐	☐	☐	☐	☐
Check blowdown operation	☐	☐	☐	☐	☐	☐	☐	☐

and process systems. Additional items may be required for boilers in applications that have unusual operating conditions.

BOILER WATER TREATMENT

One of the most effective measures to improve operating efficiency of a boiler system is a water treatment program. By reducing the level of scale and corrosion within a boiler, a good water treatment program improves the operation of not only the boiler itself, but also the steam distribution system and the condensate return system. Additionally, it contributes to the safe operation of the system as well as prolonging the life of system components.

Water treatment is a necessity as a result of impurities in the boiler makeup water, including hardness, silica, dissolved oxygen, ammonia, chlorine, chloride, sulfates, and iron. Within the boiler, these impurities become concentrated as steam is generated and form scale on boiler heat transfer surfaces. This scale decreases the effectiveness heat transfer within the boiler, lowering boiler efficiency. A scale thickness of only one-thirty-second of an inch reduces boiler efficiency by approximately 4 percent. A one-sixteenth-inch layer of scale reduces boiler efficiency by 6 percent.

As the scale builds, the temperature of the boiler tubes rises, increasing the chances for a boiler tube failure. If the level of impurities from the boiler makeup water increases sufficiently, it will cause foaming within the boiler, resulting in the carryover of water droplets and contaminants into the steam system, a condition that can lead to water hammer and damage to steam pipe fittings and valves.

A lack of proper water treatment also promotes the deterioration of the steam distribution and condensate return systems. Dissolved oxygen in the makeup water results in pitting of steam and condensate lines. Bicarbonate and carbonate alkalinity in the makeup water undergo thermal decomposition in the boiler, releasing carbon dioxide into the steam system. When the carbon dioxide reaches the condensate system, it forms carbonic acid. The carbonic acid attacks condensate piping, causing a rapid thinning of the pipe wall or grooving along the bottom of the pipe. The effects of carbonic acid are the leading reason that maintenance managers abandon portions of condensate return systems.

Water treatment systems have four major elements: clarification, demineralization or softening, deaeration, and addition of amines. Clarification is the first step in the water treatment process and is used to remove the large, suspended solids from the feedwater. Mechanical means, such as settling basins and filters, are used to remove the larger suspended solids. Chemical coagulants are used to clump smaller solids into larger masses for removal.

The second step, demineralization and softening, is designed to remove dissolved salts from the feedwater that contribute to boiler scale. Silica, iron, calcium, and magnesium are among the items removed from the feedwater by demineralization and softening systems. The two most common processes used are lime-soda softening and ion-exchange. The lime-soda process uses hydrated lime to react with the soluble calcium and magnesium bicarbonates to form insoluble precipitates that can be removed from the feedwater by settling and filtration. The ion-exchange process is effective at removing ionized impurities including calcium, magnesium, iron, and manganese.

Deaeration is the process used to remove dissolved oxygen, carbon dioxide, and hydrogen from the feedwater. Each of these dissolved gases is highly corrosive to boilers, steam distribution systems, and condensate return systems. Most deaeration systems use steam to heat the feedwater. As steam is vented from the deaerator, it carries most of the dissolved gases with it, reducing concentrations in the feedwater to as low as 0.005 ppm for oxygen.

The fourth element of water treatment, addition of amines to the boiler water, is targeted to reduce corrosion in the condensate systems. Even with good water treatment that reduces the level of contaminants in the steam system to acceptable levels, corrosion will continue to take place in the condensate system, primarily as a result of oxygen and carbon dioxide present in the condensate. To further protect the condensate system piping, traps, and valves, a filming amine is added to water. Carried throughout the system by the steam, the filming amine forms a monomolecular layer on metal surfaces within the condensate system, helping to prevent contact between the corrosive condensate and the metal surface. Filming amines are effective in reducing the rate of attack of both oxygen and carbon dioxide.

Water treatment programs are site specific and must be designed around the quality of water required for the boiler, the quality of the raw makeup water, and the quantity of makeup water used by the system. The program requires the use of someone who is specially trained in boiler water chemistry. Water samples must be taken and analyzed to determine the specific contaminants that exist in the water system and their concentration. Based on the findings of the analysis, the water treatment specialist will put together a program that recommends treatments, quantities, and feeding methods. Once started, the program must include ongoing testing of both the feedwater and the condensate return water to determine the effectiveness of the treatment program, as well as to recommend modifications necessary to provide proper protection to the water system.

Most facilities elect to contract all or part of the program to outside firms specializing in water treatment. These firms perform the survey of existing system and water conditions, recommend specific water treatments based on the conditions found, provide guidelines for operators in the use of water treatment chemicals, train in-house personnel in water testing procedures, review the periodic water tests, and recommend changes in the chemical treatment program as required.

An important element of boiler water treatment that is often overlooked by maintenance managers is taking steps to minimize the quantity of makeup water used by the boiler system. Since makeup water is the major source for impurities that enter the steam system and create the need for water treatment, the top priority for any water treatment program must be to reduce the need for makeup water. Decreased makeup water requirements reduce boiler water costs, lessen the quantity and cost of chemicals required to treat the water, and make the water treatment system simpler to operate.

Makeup water requirements can be reduced by performing routine and preventive maintenance: Return as much condensate as possible. Regularly check both the steam and the condensate system for leaks. Inspect valve stems, pump seals, relief valves, heat exchangers, and condensate receivers for leaks. The resulting decrease in operating costs more than pays for the expense of the maintenance program.

Install a meter on the makeup water line so that makeup water use can be monitored, preferably daily. Any sudden increases in the demand for makeup water are indications of leaks within the steam or condensate systems. A sudden decrease in makeup water demand may be a sign that a steam coil is leaking, allowing untreated water to enter the condensate system. In new systems, excluding the water used for boiler blowdown, the quantity of makeup water used by the system should be less than 5 percent of the steam generated by the system. In older facilities, the goal should be to reduce makeup water requirements to below 10 percent.

The performance of the water treatment program must be regularly evaluated by sampling boiler system and makeup water. Changes in water samples will indicate modifications that must be made to the treatment program. The annual inspection of the boiler will also provide feedback on the effectiveness of the program in keeping boiler water surfaces clean of scale and corrosion.

HOW TO CONTROL BOILER BLOWDOWN

Even with an effective boiler feedwater treatment program, there will be some level of dissolved solids in the feedwater. When a boiler is first filled and started, the concentration of dissolved solids in the feedwater and the boiler are equal. Most of these solids are not carried over into the steam, but remain in the boiler itself. As the boiler operates, and requires additional feedwater to make up for system losses, more solids are introduced into the boiler, increasing their concentration in the boiler. If the concentration is allowed to become high enough, scale can form on the boiler's heat transfer surfaces and sludge can accumulate in the bottom of the boiler. Both contaminants decrease the boiler's operating efficiency, increase the rate of corrosion of steel surfaces, contribute to boiler tube plugging, and can result in tube failure.

To keep the concentration of dissolved solids within an acceptable range, operators discharge a portion of the boiler's water and purge the highly concentrated impurities. This blowdown is accomplished at two different areas within the boiler. Skimmer blowdown draws boiler water from several inches below the top of the boiler's water level where the concentration of contaminants is the highest. Bottom blowdown draws water from the bottom of the boiler's mud drum to remove the sludge that accumulates there. While skimmer blowdown can be automated, bottom blowdown is a manual activity that is typically performed daily. Of the two types of blowdown, skimmer blowdown accounts for nearly 90 percent of the blowdown requirements for most boiler installations.

A system's blowdown rate is defined as the ratio of the quantity of water blowndown from the system to the total quantity of boiler feedwater. Depending on the quality of the system's makeup water and the quantity of makeup water required, the blowdown rate may range as high as 10 percent. For example, the maximum permissible concentration of solids in a boiler operating at 300 psig or lower is 3,500 ppm. If the boiler feedwater (condensate return plus makeup water) contains 350 ppm total solids, the blowdown rate required to maintain the boiler within the recommended operating range would be 10 percent. This 10 percent blowdown rate translates to a seasonal energy loss for the boiler of approximately 3 to 4 percent.

Blowdown represents a significant energy loss for the boiler. The water that is discharged during blowdown is already heated to the boiler's operating temperature and is discharged to the sewer. While heat recovery equipment can be installed to transfer some of the energy from the blowdown to the incoming feedwater, the first priority should be to ensure that the blowdown rates are properly adjusted to match boiler requirements. If the blowdown rate is too high, the water treatment chemical concentration in the boiler will be too low to be effective and energy efficiency will be low. If the blowdown rate is too low, the concentration of solids in the boiler will increase and will promote the development of sludge and scale within the boiler, reducing energy efficiency. Therefore, while any quantity of blowdown decreases the energy efficiency of the boiler, energy losses from blowdown can be minimized by matching the rate of blowdown to the boiler's requirements. Use Figure 7.3 to determine the percentage blowdown required for boilers operating at pressures of 300 psi or less based on the concentration of solids in the feedwater.

The first defense against excessive blowdown losses is to reduce the need for makeup water. Makeup water is the source for most of the impurities that must be removed through the use of boiler blowdown. Therefore reducing the need for makeup water will reduce the volume of blowdown required to maintain water quality within the required limits.

Blowdown losses also can be controlled by properly setting the blowdown rate. Too often, controls are out of calibrations or operators set the blowdown rate too high, both of which result in high energy losses. For example, a reduction in the blowdown rate from 20 to 10 percent in a 200 psig boiler producing 500,000 pounds of steam per day results in a daily savings of nearly $100 at an energy cost of five dollars per million Btu.

If the skimmer blowdown system is manually controlled, water samples should be taken at least once per shift and tested for the concentration of dissolved solids using a conductivity test. Water samples should not be taken while the boiler is undergoing a change in load to ensure the accuracy of the sample. Based on the results of the testing, the blowdown rate should be adjusted. Large boilers should be tested more frequently, or should be fitted with an automatic blowdown system. **Rule of Thumb:** A properly calibrated automatic blowdown system will reduce blowdown rates annually by 15 to 20 percent over the rates achieved by manual controls.

Automatic skimmer blowdown systems can maintain the quality of the water within the recommended range as long as they are functioning properly. At least once per day, a water sample should be taken from the blowdown system and tested for the concentration of dissolved solids using a conductivity test. Test results will indicate if the system is operating properly.

Figure 7.3
Boiler Blowdown Requirements

Percent Blowdown

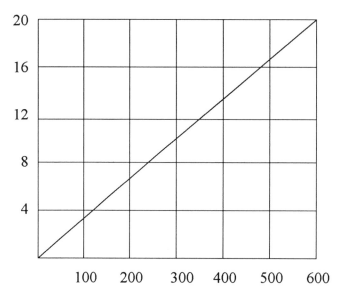

Feedwater Solids Concentration (ppm)

HOW TO CONTROL EXCESS COMBUSTION AIR

One of the most cost-effective means of improving operating efficiency is by controlling the amount of combustion air supplied to a boiler. The quantity of air required by a boiler depends on the boiler's firing rate and the type of fuel used. Too low a flow rate of combustion air results in incomplete combustion of the fuel leading to the creation of smoke, increased levels of carbon monoxide, and sooting of heat transfer surfaces. Too high a rate and the energy efficiency of the boiler will decrease due to losses up the boiler's stack resulting from the boiler's having to heat the extra air.

All boilers require more combustion air than the theoretical minimum to ensure complete combustion of the fuel. The efficiency drop-off for too little combustion air is much more pronounced than for too much air. Additionally, sudden load changes on the boiler might result in conditions where low combustion air supply rates might result in boiler explosions. Therefore boilers are commonly operated with excess combustion air levels of 25 to 50 percent.

Few boilers require high excess combustion air. Most natural gas-fired boilers require no more than 10 percent excess air. Number 2 oil-fired boilers require no more than 12 percent excess air. Number 6 oil-fired units operate efficiently with 15 percent excess air. Any higher level of excess air results in decreased efficiency. *Rule of Thumb:* each percentage point reduction in the level of excess air supplied to a boiler results in a one-half to one percent improvement in efficiency. For example, reducing the excess air supplied to a 40,000 lb/hr boiler by 5 percent will result in an annual savings of more than $120,000 for fuel at a rate of five dollars per million Btu.

The quantity of excess air being supplied to a boiler can be estimated by measuring either the oxygen or the carbon dioxide concentrations in the boiler's exhaust flue gases. Use Figure 7.4 to estimate the percentage of excess air based on the measured level of oxygen in the flue gas. Use Figure 7.5 to estimate the percentage of excess air based on the measured level of carbon dioxide in the flue gas.

Boiler combustion air is controlled through either automatic or manual controls. While automatic controls provide better regulation of excess air levels, both types of controls require constant monitoring and adjustment to ensure proper control action. On small boilers, flue gases should be sampled at least once per week and necessary adjustments made to the combustion air controls. On larger boilers, flue gases should be sampled at least once per shift.

BURNER MAINTENANCE

Maintenance of the burner system in a boiler must focus on both safety and efficiency. Burner performance is impacted by changes in combustion air temperature, barometric pressure, fuel pressure, and in the case of oil fuels, fuel temperature. Therefore to obtain the best performance out of the burner system and enhance the safe operation of the boiler, the burner and fuel systems must be properly maintained.

In natural gas-fired boilers, gas pressure is critical to proper burner operation and

Figure 7.4
Boiler Flue Gas Oxygen Content Resulting from Excess Combustion Air

Flue Gas O_2 (%)	Excess Combustion Air (%)
1.0	3.5
2.0	7.0
3.0	15
4.0	20
5.0	28
6.0	35
7.0	50
8.0	65
9.0	80
10	96
12.5	150
15.0	180
17.5	220
20.0	275

Figure 7.5
Boiler Flue Gas Carbon Dioxide Content Resulting from Excess Combustion Air

	Excess Combustion Air (%)		
Flue Gas CO2 (%)	Natural Gas	#2 Oil	#6 Oil
6	100	-	-
7	55	-	-
8	35	110	150
9	18	90	120
10	7.5	55	80
11	2	38	60
12	-	25	40
13	-	15	25
14	-	8	15
15	-	3	8.5
16	-	-	4.5
17	-	-	1.0

efficient combustion. Variations in gas pressure may cause over- or underfiring of the boiler. Overfiring may result in flame impingement on the walls of the boiler, insufficient air for proper combustion, and accumulation of soot on the interior surfaces. Underfiring may cause flame instability and poor combustion efficiency due to high levels of excess air. Both conditions result in abnormal gas circulation patterns in the boiler's heat transfer region, further decreasing operating efficiency.

Gas pressure should be checked and recorded at regular intervals. During steady-state operations, pressure should remain constant. During firing rate changes, pressure may vary slightly but should not oscillate. Abnormal variations can be the result of a faulty pressure regulator or debris in the gas line.

In oil-fired boilers, regulation of both the temperature and the pressure of the oil are critical to safe and efficient operation of the boiler as both impact the ability of the oil to properly atomize and burn completely. Variations in either temperature or pressure result in flame failure, increased stack emissions, fuel-rich combustion, and soot accumulation within the boiler.

Regularly check and record the temperature and pressure of oil being fed to the boiler. Variations are signs that there may be problems with the pumps, heaters, strainers, or pressure regulator. Each of these components should be inspected and overhauled at least once per year.

One of the easiest checks on the operation of the burner is a visual inspection of the flame. Changes in the shape, color, or sound of the flame are an indication of burner problems ranging from changes in the fuel pressure to dirty or worn nozzles. Binding or slipping fuel and air linkages result in changes in the air–fuel ratio within the boiler, changing the color, shape, and sound of the flame.

BOILER PREVENTIVE MAINTENANCE ACTIVITIES

One of the most important elements in keeping boilers operating at their peak efficiencies is a program of regularly scheduled maintenance and inspections. Various maintenance tasks and inspection activities are typically performed on a daily, weekly, and monthly basis. Once a year, the boiler is shut down and opened up for a detailed internal inspection. During that annual inspection, additional maintenance activities are performed.

The types of maintenance activities that are required, the items to be inspected, and their related frequencies are relatively independent of the type of boiler and operating pressure. However, specific requirements of a particular application may require additional activities and inspections, or may require more frequent implementation. For example, many of the maintenance activities and inspections that are performed on a daily basis in boilers serving single buildings often are performed once per shift or once per hour in boiler plants serving multiple buildings. The high cost of operating these boilers justifies the additional maintenance and inspection expenses.

Figure 7.6 lists recommended daily and weekly maintenance and inspection activities for boilers. Figure 7.7 lists maintenance and inspection activities that are to be performed on a monthly basis. Figure 7.8 lists annual maintenance and inspection

Figure 7.6
Daily and Weekly Boiler Maintenance Activities

1. Building/facility: _____

2. Room #: _____   3. Boiler #: _____

4. Dates: _____

Activity	Day of Week						
	1	2	3	4	5	6	7
Complete boiler operating log							
Check general operating conditions							
Check water level							
Test low water cutoff							
Manually blowdown boiler bottom							
Inspect automatic surface blowdown							
Blowdown water column							
Visually check combustion							
Inspect burner controls							
Inspect burner operation							
Check water treatment system							
Check auxiliary equipment							
Check stack temperature							
Check stack oxygen/carbon dioxide							
Test relief valve							
Check water level control							
Check air and fuel linkages							
Test safety and interlock controls							
Test indicator lights and alarms							

Figure 7.7
Monthly Boiler Maintenance Activities

1. Building/facility: _____

2. Room #: _____   3. Boiler #: _____

4. Dates: _____

Activity	Month					
	1	2	3	4	5	6
Inspect burner						
Check tight closing of fuel valve						
Inspect for gas/oil leaks						
Inspect for flue gas leaks						
Inspect for hot spots						
Inspect for air leaks						
Blow down all control lines						
Zero meters and draft gauges						
Inspect all belts						
Inspect all pump couplings						
Inspect all auxiliary equipment motors						
Check blowdown equipment						
Check water treatment equipment						
Inspect for leaks, noise & vibration						

Figure 7.8
Annual Boiler Maintenance Activities

1. Building/facility: _____

2. Room #: _____   3. Boiler #: _____

4. Dates: _____

	Completed	Comments
Inspect and clean waterside surfaces	☐	_____
Inspect and clean fireside surfaces	☐	_____
Eddy current test tubes	☐	_____
Inspect all refractory materials	☐	_____
Inspect all manhole gaskets for leaks	☐	_____
Inspect and test all valves and cocks	☐	_____
Overhaul boiler auxiliaries	☐	_____
Clean and rebuild low-water cutoff	☐	_____
Rebuild pressure relief valve	☐	_____
Test operation of all safety devices	☐	_____
Check operation of hydraulic and pneu-matic valves	☐	_____
Recalibrate all operating controls	☐	_____
Overhaul feedwater pumps	☐	_____
Change turbine and reduction gear oil	☐	_____
Clean condensate receiver	☐	_____
Overhaul oil preheater	☐	_____
Clean/replace oil pump strainers and filters	☐	_____
Inspect electrical terminals	☐	_____

activities. These lists of maintenance activities are considered to be a minimum listing for boilers in typical applications and will have to be expanded for unusual circumstances. Additional activities may be required for some boilers depending on the design of the boiler and the particular application. For additional information on maintenance requirements, contact the boiler's manufacturer.

STEAM DISTRIBUTION SYSTEM COMPONENTS AND HOW TO MAINTAIN THEM

There are three major components to a steam distribution system: steam piping from the boiler to the heating load, steam traps, and condensate return piping. The proper operation of each component is essential to the efficient operation of the steam system. Proper operation requires ongoing inspection and maintenance of all components. Unfortunately, these components are often ignored even when they are obviously malfunctioning. Steam leaks, blowing traps, and dumped condensate are all signs of a steam distribution system that has been ignored and not properly maintained.

Steam Piping

There are two major activities required in order to maintain the efficiency of the steam piping system: correction of steam leaks and repair of steam line insulation. Both are costly losses, yet both are frequently ignored. Surveys of above ground steam piping routinely identify numerous leaks, wet or damaged insulation, and missing insulation. Infrared surveys of underground steam lines frequently show high heat loss rates due to damaged or ineffective insulation. In many cases, the heat loss is so high that it is easy to trace the exact route of the steam line by walking the path of melted snow or dead grass even though the line may be buried three feet or more below the surface.

Steam leaks are often ignored because their cost is not understood. Most leaks at pipe flanges or valve stems are generally considered by many maintenance managers to be too small to be significant. While a steam leak is visible and usually audible, it is overlooked because its repair typically requires a shutdown in steam service and removal and repair or replacement of the leaking component. The costs of the shutdown and repairs simply are thought to be too costly.

Steam leaks are very costly to the operators of steam systems. For example, a one-eighth-inch diameter hole in a 100 psig steam distribution system loses steam at the rate of more than 53,000 pounds per month. With time, the escaping steam cuts through the surrounding metal, enlarging the hole and increasing the rate of steam loss. When the hole expands to three-sixteenths of an inch, its loss rate is more than 115,000 pounds of steam per month. A one-quarter inch diameter hole looses more than 200,000 pounds of steam per month. At a steam cost of seven dollars per thousand pounds, a one-eighth diameter hole wastes more than $370 in steam per month. The cost of a one-quarter inch leak is more than $1,400 per month.

Damaged or missing steam pipe insulation also causes significant energy loss in steam systems. Unlike steam leaks, damaged and missing insulation is not readily detected. Insulation may become saturated with water and gave little or no outward indication that it is no longer effective. Even areas where insulation is missing may not be obvious. The longer an area goes without insulation the more it becomes accepted as the norm.

Properly installed insulation in good condition will reduce uninsulated steam pipe heat loss by 90 to 95 percent. For example, an uninsulated one-inch diameter pipe carrying steam at 100 psig will lose heat at the rate of more than 4,000,000 Btu per foot of pipe. At a steam cost of seven dollars per thousand pounds, properly insulating that one-foot section of pipe will save more than $2,000 per year.

Nearly every steam system will have numerous small leaks and areas where insulation is damaged or missing. Although each leak and poorly insulated area wastes energy, it is their large numbers typically found in a steam system that make their cost significant. In large distribution systems it is not uncommon to find that 10 percent or more of the total steam generated by the system is lost through leaks or poorly insulated piping.

In order to identify the location of leaks and areas where insulation is in need of replacement, it will be necessary to survey the entire steam system. All leaks, regardless of size, must be identified and scheduled for repair. In some cases, repairs will have to be postponed until the system can be shut down for annual maintenance. To facilitate repairs, stock all repair components well in advance of the shutdown.

The steam system survey also is used to identify areas where insulation is in need of repair or replacement. Particular attention should be paid to areas that traditionally have been left uninsulated, including valves, elbows, and other pipe fittings. The cost of the heat lost through these fittings and components typically is sufficient to recover the cost of the insulation installation in less than six months. A side benefit of repairing steam pipe insulation is that it may help to lower the temperature and humidity in some mechanical equipment rooms.

Once the initial survey has been completed, it should be repeated on an annual basis. Leaks and damage can occur at any time, and need to be identified and corrected as quickly as possible to minimize both energy loss and further damage to system components.

Underground steam lines can be a major source of energy loss within a distribution. Often these losses go undetected for years simply because they are out of sight. The most common cause of damage to the insulation on underground steam piping is ground water that has either penetrated the outer casing of the line, or has entered the line through a flooded manhole. In both cases, the water eventually floods the entire outer case of the line, destroying the effectiveness of the insulation and promoting the corrosion of the line itself. The heat loss through sections of underground steam lines with damaged insulation is typically six to eight times the rate of heat loss through piping with sound insulation.

There are two steps to minimizing losses from underground steam lines: keep water out of manholes, and perform infrared surveys of the lines. Since manholes are a major source of water that can damage the insulation in steam lines, it is important to keep water from accumulating in the manholes. When a manhole floods, water can enter the conduit surrounding the steam line through the conduit's drain line. Eventually, it will corrode the line's outer jacket from the inside, allowing even more groundwater to enter the line's insulation.

Inspect manholes for water accumulation. Ones where water is found to be accumulated must be pumped out and the source of the water identified. If the manhole cannot be sealed to reduce or eliminate the entrance of groundwater, pumps will have to be installed to carry the water away before it can damage piping and insulation on steam lines within the manhole and steam lines connected to the manhole. After the initial survey, it will be necessary to inspect all manholes on a monthly basis to ensure that they are remaining free of water.

The infrared survey will identify areas where underground pipe insulation has deteriorated, resulting in high rates of heat loss. These surveys are typically performed from aircraft at night, using a special high-resolution infrared camera. The images produced by the cameras will not only help to identify areas having high heat losses, but also will quantify the heat loss rate. Information provided by the infrared survey is the first step in identifying areas requiring maintenance activities or possible replacement. Once areas with high heat loss rates have been identified, several test holes will have to be made to uncover small portions of the line so that its condition may be better evaluated for repair or replacement.

Steam Condensate Systems

Steam condensate systems are the most ignored component in the steam system. The corrosive nature of condensate, particularly in systems with an inadequate water treatment program, results in the need for frequent maintenance. As a result, many facility managers have simply abandoned large portions of their condensate return systems rather than expend the money necessary to keep them operating properly. While the ideal condensate return system would return 100 percent of the condensate produced by the steam system, most systems fall far short of this goal. Many return less than 50 percent.

Returning condensate to the boiler reduces operating costs in several ways. Condensate on average contains 15 to 20 percent of the heat content of the steam produced by the system. By returning condensate to the boiler, this heat energy is recovered. Recovering condensate reduces the quantity and cost of makeup water required by the boiler. Since condensate is relatively free of the contaminants found in makeup water, the quantity and cost of water treatment chemicals required by the system is reduced. Reducing the quantity of makeup water required also reduces the rate of blowdown required, further increasing system efficiency and decreasing makeup water costs.

The primary goal of a condensate system maintenance program is to return as much contaminant-free condensate as possible to the boiler. Start with the boiler log. Included in the data recorded in the daily log is the quantity of steam generated and the quantity of boiler makeup water use. Any significant change in makeup water use with respect to steam generation is an indication that condensate is leaking out of the system, or that untreated water is leaking into the system, mixing with the condensate.

If maintenance on the condensate system has been ignored or inadequate for a long period of time, most likely it will be necessary to replace major portions of the system. All portions of the condensate system should be inspected and surveyed for damage. Areas that have been abandoned or that are not properly operating should be identified for repair or replacement as quickly as possible. Inspect damaged areas to determine the cause for the failure. If piping is badly pitted, the condensate contained oxygen. If the pipe wall is badly thinned, or if there is deep grooving in the bottom of the condensate piping, the condensate contained high levels of carbon dioxide that formed carbonic acid. Both of these conditions are caused by improper or inadequate water treatment and must be corrected before any repairs or piping replacements are made in the condensate.

If large sections of the condensate return system need replacement, establish a phased program to spread the program cost over several years. The priority for replacing a particular section can be established based on the volume of condensate that it would return to the boiler.

Once the system has been upgraded to return as much condensate as possible, the system's operation must be constantly monitored. Collect samples of condensate daily and test for ionized mineral content, hardness, and total suspended solids, all three of which should be low for clean condensate. Inspect the condensate sample for color, odor, and turbidity. Any changes in the sample from normal are indications that the water treatment program is not operating properly or that there are leaks in the return system. Use Figure 7.9 to record and track daily condensate readings.

In addition to daily testing, the complete condensate system should be inspected once a year. All sources of air and water leaks must be identified and corrected. Equipment connected to the system, such as heat exchangers and heating coils, should be examined for leaks. If possible, the system should be opened at several locations and the piping inspected for signs of corrosion and pitting.

Steam Traps

An important component of the steam condensate system that plays a major role in the overall efficiency of the system is the steam trap. Its main purpose is to function as an automatic valve that allows condensate to pass while retaining steam. Some trap designs also function to separate air from the system to reduce the levels of both oxygen and carbon dioxide in the condensate return system.

Traps malfunction in one of two ways—they can fail open, allowing steam to pass through the trap into the condensate system, or they can fail closed, allowing condensate to build up in the steam system. A trap that has failed open wastes steam, decreasing the efficiency of the system. A trap that has failed closed is a safety hazard. Trapped condensate is accelerated by the steam to high velocities and can impact pipe fittings, elbows, and valves with sufficient force to break them. In spite of the important function performed by steam traps, they are often ignored.

Figure 7.9
Condensate Monitoring

1. Building/facility: _____

2. Room #: _____ 3. Boiler #: _____

4. Dates: _____

Activity	Data						
Date							
Time							
Temperature							
Color							
Odor							
Turbidity							
Hardness							
Ionized mineral content							
Total suspended solids							

There are several factors that contribute to the ignoring of steam traps by maintenance personnel. One of the main factors is that their operational components are hidden from view. One cannot tell a trap has failed simply by looking at it. As a result, they are ignored until their failure results in occupant discomfort or damage to the steam or condensate system.

Another factor contributing to the ignoring of steam trap maintenance is a misunderstanding of the importance of having traps operate properly. Trap repair and replacement is a labor intensive activity and therefore frequently deferred in favor of more obvious and easily understood maintenance tasks. Few maintenance managers realize that a single failed trap on a 30 psig steam line will lose more than $400 in steam per year.

Surveys of facilities routinely show that anywhere from 10 to 50 percent of the installed steam traps are not properly operating, creating both energy waste and safety hazards. Implementation of a comprehensive steam trap maintenance program will not eliminate all trap problems simply because of the large number of traps installed in even moderately sized facilities. The program, however, can reduce energy losses through failed traps. The resulting improvement in operating efficiency of the total system averages between 5 and 10 percent.

The rate at which steam that is lost through a trap depends on the operating pressure of the steam line to which the trap is connected, and the size of the orifice in the trap. For example, a typical trap connected to a 30 psig steam line may have a one-eighth or a one-fourth-inch orifice, depending on the heating capacity of the device served by the trap. The trap with the one-eighth-inch orifice will pass steam at the rate of 12 pounds per hour. The trap with the one-fourth-inch orifice will allow more than 45 pounds of steam to pass every hour. If even 10 percent of the traps in a steam system have failed, the energy losses will be very significant, particularly if not all of the condensate is returned to the boiler.

Although steam traps are fairly simple in their design, their high failure rate can be attributed to the environment in which they must operate. Oxygen and carbon dioxide common contaminants in condensate, severely corrode the internals of steam traps. Thermal stresses resulting from the constant cycling of the traps contribute to failure of the trap's internal components. Dirt, scale, and other solid contaminants present in the steam can clog or erode seats within traps. The flow of a mixture of steam and hot condensate erodes internal passages in the steam trap. The result is an atmosphere that rapidly destroys steam trap operations.

Steam Trap Maintenance Program

The only means for ensuring that the steam traps within a system are functioning properly is to implement a regular trap inspection and repair/replacement program. The program starts with a survey of all steam traps within the facility. This survey locates all traps and catalogues their sizes, types, and models. Because there are so many traps in a typical steam system, and many of the traps may be hidden from view, it is critical that those completing the initial survey take the time both to find all of the traps and to provide sufficient information on their location so that others may find them at a later date. Most surveys would benefit from the development of a mapped route that identifies both the location of the traps and the recommended route to be followed when inspecting them.

In addition to locating and cataloguing traps, the survey must also test their operation. There are three methods for testing the operation of steam traps: visual observation, acoustic monitoring, and temperature measurement.

☐ *Method 1*. Visual observation is the easiest and most cost-effective method of testing trap operation. If the condensate line immediately downstream from the trap can be opened, the discharge from the trap can be examined. A properly operating trap will intermittently discharge condensate and a small amount of flash steam. If there is no discharge, the trap has failed closed. If the trap discharges a continuous stream of steam, the trap has failed open. Testing by visual observation can be performed only when there is a way to discharge the trap to the atmosphere.

☐ *Method 2*. Acoustic testing of traps involves listening carefully to the operation of the trap through an industrial stethoscope or an ultrasonic detector. Properly operating traps will cycle over time, resulting in changes in the sounds they emit. Failed traps will either emit a constant sound, or no sound at all. With training and practice, maintenance personnel will recognize the differences and be able to quickly and accurately diagnose failed traps.

☐ *Method 3*. Measuring the temperature difference across a steam trap is also a way to identify traps that have failed. Properly operating traps display a significant temperature drop. Traps that have failed open show little temperature drop. Traps that have failed closed show a large temperature drop, with the outlet side being close to ambient temperatures. Temperature readings should be taken at locations two to three feet on either side of the trap to reduce faulty readings resulting from thermal conduction across the trap. Commonly used temperature measurement equipment includes infrared meters and contact thermocouples.

The large number of traps found in most steam applications will make completion of the initial survey a major undertaking. If your facility lacks the resources to rapidly complete the survey, it can be broken down into several phases. Include all traps on steam mains, high-pressure distribution piping, and drip legs in the first phase of the program. These traps operate at the highest pressure and have the potential for the greatest losses in the system. Proper operation of these traps is also critical to the safe operation of the steam system.

The next phase of the trap maintenance program includes medium- and low-pressure, high-volume traps typically found on large heating coils, particularly if steam to these coils is not regulated by a steam control valve. Although these devices operate at lower pressures than steam mains, the high volume of steam flow through them increases the potential for losses. The final phase includes all remaining traps in the system. These devices typically operate at low pressure and have a lower but still significant potential for losses.

Once the initial survey has been completed and the failed traps replaced, it will be necessary to repeat the survey at regular intervals. Steam traps are relatively short life components of the steam system, requiring frequent replacement. Under good operating conditions, traps can be expected to last an average of only five to ten years. As the level of contaminants in the system increases, that life expectancy decreases significantly. Therefore, it is recommended that the trap testing be repeated every one to two years. With the initial trap survey completed and the location of the traps known, these additional rounds of testing will not be so labor intensive. Maintain records of traps that have failed or have required maintenance. These records will help in identifying recurring problems. As experience is gained with the trap maintenance program, it may be possible to alter the inspection schedule.

TWO OPERATING PRACTICES FOR ENERGY CONSERVATION

There are a number of operating practices that can be implemented to improve the seasonal efficiency of the boiler system. Most are simply attention to detail in the day-to-day operation and management of the facility. Some involve developing policies to ensure future efficient operation, such as requiring any connection to the steam system to also include the proper additions for returning condensate to the boiler.

Two operating practices that will improve seasonal efficiency of the system are proper load management and lowering operating pressures. Before implementing one or both of these practices, operators must make certain that they are compatible with the facility.

Load Management

Load management is an effective tool to reduce boiler energy use in facilities with multiple boilers, such as central boiler plants serving multiple buildings. Properly implemented, it typically produces an annual savings of 1 to 2 percent. Load management produces its savings by taking advantage of the fact that boiler efficiency varies not only with the design of the boiler and the fuel it uses, but also with its condition and load.

Under load management, operators use the most efficient boilers available and operate each boiler at a load as close as possible to its peak efficiency. As the load changes on the plant, it will be necessary to both distribute the load properly between the currently firing boilers as well as to see that the most efficient boilers are on line.

In order to implement load management effectively, it is necessary to determine the efficiency curves for each boiler in the plant. Measurements will have to be taken for various loads for each unit, and the results plotted into an efficiency curve. For most boilers, peak operating efficiency occurs at a load below full load.

Once the performance data for boilers are known, operators must constantly monitor load and operating conditions to ensure that the most efficient combination of boilers is on line for the load conditions and that the total plant load is properly distributed across all on-line boilers.

Lowering Operating Pressure

Lowering the operating pressure of the system offers several ways in which energy use can be reduced. Within the boiler itself, lowering the steam pressure will result in a decrease in stack temperatures and a resultant improvement in the efficiency of the boiler. For example, lowering the operating pressure of a system from 150 to 100 psig will result in an increase in boiler efficiency of approximately 1 percent.

Additional savings result from decreased rates of loss throughout the steam system. Although the maintenance program should strive to minimize the number of leaks in the system, all steam systems have leaks. Lowering the operating pressure of the system will decrease the rate of loss from these leaks. For medium, multi-building systems in good operating condition, typical energy improvements range from a 1 to 2 percent increase in system efficiency due to lower loss rates through leaks.

Finally, lower operating pressures decrease conduction loss rates from steam piping. While the savings produced in piping whose insulation is in good condition are rather small, more significant savings can be produced in piping sections where the insulation is damaged. For example, the conduction losses from underground piping with damaged insulation are reduced by lowering the system operating pressure.

Reducing system operating pressure is not feasible in all steam systems. In large systems or systems with only marginal capacity, there may not be adequate pressure to deliver steam to all points in the system. In some systems, it may not be possible to reduce operating pressures year round due to the demands placed on the distribution system. It may, however, be possible to reduce operating pressure during the summer when heating loads are greatly reduced.

Operating at lower pressures may require additional modifications. Flow devices may require recalibration for proper operation at the lower pressures. Boiler controls may require modification to ensure no moisture is carried over into the steam system from the steam drum.

SUMMARY

Boiler systems are large energy users in facilities. By targeting maintenance efforts toward the components of the boiler system, you can achieve very cost-effective energy savings. Increases in energy efficiency resulting from a systematic and planned maintenance program can reduce energy use from 10 to 30 percent, with most programs achieving a payback in months, not years. Additional savings are achieved through reduced boiler water and chemical requirements. The side benefits of the maintenance program include enhanced system safety, more reliable operation, and longer system life.

Chapter 8

Air Handling Systems

This chapter presents information on maintaining building air handling systems. The information presented in this chapter may be applied to most types of air handlers commonly found in facilities. The maintenance activities and their recommended frequencies listed in this chapter represent the minimum recommended level of maintenance required by these systems and their components. Special systems or systems supporting critical applications will require additional activities and will require more frequent attention. If additional information is required in establishing a maintenance program for a particular piece of air handling equipment, it is recommended that you contact the equipment's manufacturer.

For the purposes of this chapter, the components of an air handling system are defined to include the fans, dampers, heating and cooling coils, controls, and the ductwork typically found in air handling units.

AIR HANDLING SYSTEM ENERGY MAINTENANCE CONSIDERATIONS

Air handling systems represent a major investment for building owners in terms of both the initial purchase and the ongoing operating costs. Additionally, the replacement or major overhaul of an existing air handling system causes a loss of service and major disruption for building occupants, usually sufficient to force a move to temporary facilities. Therefore, in order to gain the most from the initial and ongoing investment in air handling systems, it is essential that the units be properly maintained.

When maintenance of air handlers is neglected, a number of things can go wrong that will adversely impact the comfort of building occupants, the quality of the air being supplied to the conditioned spaces, and the operating efficiency of the air handler. Heating and cooling coils can become partially blocked by dirt and debris from the air stream, reducing their capacity and efficiency. Temperature control systems can become fouled with dirt and water, causing the system to operate above or below the desired temperature or air flow rate resulting in wasted energy or uncomfortable space conditions. Outside air dampers can be improperly positioned, allowing too little or too much ventilation air into the building, wasting energy, or causing poor air quality. Variable air volume controls can stick, resulting in excessive air flows, poor temperature control, and wasted energy.

Good maintenance of building air handling systems will result in reduced energy use, longer equipment life, less frequent breakdowns, fewer interruptions of service for occupants, improved comfort, and enhanced indoor air quality. In spite of these advantages, most air handling systems today are not included in comprehensive preventive maintenance programs. The typical facility maintenance program for air handlers includes only the most basic of maintenance tasks: lubricating select components and changing filters. All other maintenance activities are performed only after something has failed or an occupant has complained.

Breakdown maintenance remains the preferred approach to maintaining air handling systems because it is believed to be the lowest cost maintenance approach to these systems. But breakdown maintenance of air handlers has a number of hidden costs that are often overlooked when comparisons are made to comprehensive preventive maintenance programs. One of the most significant of these hidden costs is caused by creeping inefficiency.

As air handling system components age, changes take place within the system. Controls go out of calibration. Dampers fail to properly position or seal. Coils become clogged with dirt. Air flow rates change. The result of these changes is a slow decline in system operating efficiency. Routine maintenance, such as keeping motors and bearings lubricated and changing filters, does little to keep components and systems operating at their peak efficiencies. Creeping inefficiency can be countered only through a comprehensive maintenance program for air handling systems.

One of the most serious aspects of creeping inefficiency is that the underlying cause generally goes undetected and uncorrected. Causes, such as higher than required ventilation rates, improperly positioned or stuck outside air dampers, and poorly calibrated controls are missed more often than they are detected. Even when problems are detected, changes are frequently made to the system to compensate for the problem without actually correcting it. Systems may be operating very inefficiently, but as long as they can supply sufficient heating and cooling to the conditioned space most of the time, their inefficiencies and the causes for those inefficiencies will remain undetected and uncorrected.

On the surface, comprehensive maintenance may appear to be more costly than breakdown maintenance. It requires additional up-front funding for both materials and labor. But in exchange for these costs, breakdown repair costs are reduced. More importantly, operating costs, including system energy use, are reduced. In nearly all applications, the additional funds required to implement the comprehensive maintenance program are more than offset by the savings from these two areas alone. And these savings do not take into consideration other hidden costs of poor air handler maintenance, such as irate tenants, lost work time, and damage caused by frozen piping.

ELEMENTS OF THE AIR HANDLING SYSTEM AND THEIR MAINTENANCE REQUIREMENTS

While air handling systems come in a wide range of designs and sizes, most make use of the same basic components. In order for the air handler to operate as efficiently as possible, each of the system's components must be maintained properly. Maintenance of these components will vary only slightly from system to system as dictated by the particular design and requirements of the application.

The air handler maintenance program must focus on several key areas to be effective. These areas include the unit's fan, the rate at which ventilation air is supplied to the conditioned spaces, the unit's heating and cooling coils, the air filtration system, and the setup and operation of the unit's control system. The maintenance requirements for each of these components will be discussed in the following sections. Suggested maintenance activities for all components will be provided in a later section of this chapter.

Fans

The heart of the air handling system is the fan. Fans provide the work necessary to move the air through the filtration system, across the heating and cooling coils, through the ductwork, and to the conditioned space. Because they are such an obvious and important component in the air handling system, they generally receive at least a minimum level of preventive maintenance: inspection of the motor and drive components, and lubrication of motor and fan bearings. But fans require additional maintenance activities if they are expected to operate properly and efficiently over their rated life.

Fans in air handling systems are designed to deliver a specific quantity of air to the conditioned space. The rate at which air is supplied by a particular fan varies according to a set of formulas known as fan laws. These laws dictate a number of performance variables, including static pressure, the flow of air in the system, and the horsepower requirements of the fan. Variations in one performance variable will produce variations in other performance variables, variations that can be determined by applying the fan laws.

While there are a number of fan laws, there are three that are of particular importance in maintaining building air handling systems and managing fan energy use. The first of these laws defines the relationship between air flow in the system and the speed of the fan. If all other factors remain unchanged, the air supplied by the fan to the air handling system varies directly in proportion to the speed of the fan.

The second fan law defines the relationship between fan speed and the static pressure in the air handling system. Assuming all other factors in the system remain the same, a slight increase in fan speed results in a large increase in system static pressure.

The third fan law defines the relationship between the speed of the fan and the horsepower required to power the fan. Fan horsepower requirements are related to the cube of the speed of the fan. Again, assuming all other factors remain unchanged, a slight increase in fan speed results in a large increase in fan horsepower requirements.

What these laws mean for controlling fan energy use in air handling systems is that it is very important that fan systems be operated as close as possible to design conditions. For example, according to the fan laws, a 10 percent increase in the operating speed of an air handling system fan results in an increase in the air supplied to the system of about 10 percent, while the energy use of the fan increases by nearly one-third. Similarly, a 10 percent decrease in fan speed results in a decrease in the air supplied to the system by approximately 10 percent, but a decrease in fan energy use of nearly one-third. It is this rapid decrease in fan energy use with decreased air flow that has led to the widespread use of variable air volume system designs.

Proper fan operation requires a program of ongoing inspections and preventive maintenance activities. *Rule of Thumb:* Maintenance activities are typically performed on a monthly basis, with more extensive inspection and repair activities being performed annually. The operation of all systems should be inspected at least once per week. The operation of large fan systems, particularly those greater than 25 horsepower, should be inspected once per day.

Ventilation Rates and Indoor Air Quality

Today's tighter building designs limit the amount of outdoor air that can infiltrate into the facility. At the same time, the increased use of synthetic building materials and finishes increases the quantity of contaminants generated within the occupied space. To limit the buildup of contaminants that building occupants are exposed to and to provide fresh air, building air handlers are designed to provide ventilation air. Ventilation air is air from outside the facility that is drawn into the air handling system, mixed with air that is being returned from the conditioned space, filtered, conditioned, and then supplied to the facility's conditioned space. The quantity of ventilation air that must be supplied depends on a number of factors, including the occupancy of the space, the activities that are taking place there, and the type and level of pollutants that are being produced.

Indoor air quality (IAQ) is the perception of the condition of the air and its contents within a facility. One of the most significant causes of IAQ problems are emissions from indoor contaminants, including building materials, cleaners, furnishings, standing water, damp surfaces, office equipment, processes, and combustion appliances. Other contributing factors include low ventilation rates, poor maintenance of building HVAC equipment, ineffective filtration of the air supply, and poor system design. While the causes of poor IAQ are diverse, nearly all fixes of IAQ problems involve either physical modifications to the building's air handlers, or modifications to how those systems are operated.

One of the most commonly applied fixes for IAQ is to simply increase the rate at which ventilation air is brought into the facility. Increased quantities of ventilation air help to dilute the emissions from indoor contamination sources. But increasing the rate of ventilation air does not necessarily solve the causes of IAQ problems, it simply makes them less noticeable. And increasing the rate of ventilation air increases the energy requirements for both heating and cooling.

A better approach is to determine the cause or causes of the poor IAQ and take the necessary steps to correct the problem rather than simply supplying more ventilation air. First, identify the source of the problem. If it is an accumulation of organic and inorganic fumes in the facility, identify their source and remove it from the facility. Often, substitution of one material for another will eliminate the problem.

If the IAQ problems are the result of a process or a particular piece of equipment that gives off vapors and compounds, isolate the equipment in a room that can be ventilated separately from the building HVAC equipment. The cost of installing a small, specialized ventilation system to support equipment that generates fumes is far less than having to operate a large portion of the building at a ventilation rate higher that otherwise necessary.

Another major contributor to IAQ problems is the growth of living organisms in the occupied spaces, in areas adjacent to the occupied spaces, and within air handling system itself. Identify the areas where the organisms are growing, such as on internal duct insulation, and take corrective steps to both remove the growth and prevent it from returning.

External sources also can contribute to IAQ problems. Inspect all outside air intakes to air handlers and determine if there are sources of pollutants located nearby. It is surprising how often the exhaust air from a cooling tower is located next to an air handler's outdoor air intake. Building exhaust systems, standby generator exhausts, fume hood exhausts, and even vehicle traffic are common sources for fumes that are readily brought into the building.

Finally, make certain that the system is operating as intended. Many system-related IAQ problems are the result of alterations made to the system or to the space itself. Often maintenance is not properly performed, resulting in conditions that contribute to IAQ problems. Equipment may not be operating properly. System controls may have been bypassed in an attempt to get around some other problem. Address these issues first before considering increasing the ventilation air rate.

Although ventilation air is only a portion of the total air supplied to a given area within the facility, it represents a very significant energy cost. During the heating season, ventilation air must be heated. During the air conditioning season, it must be cooled and dehumidified. Too little ventilation air results in a potential buildup of contaminants and poor indoor air quality. Too much ventilation air increases energy use.

The American Society of Heating, Refrigeration, and Air Conditioning Engineers has established a set of standards for determining the quantity of ventilation air required for a range of facilities—Standard 62-1989. Figure 8.1 lists the recommended minimum ventilation rates for a range of applications. The ventilation rates listed in the figure are for typical applications, based on the occupancy of the facility. These ventilation rates are minimum recommended values and may have to be modified for particular applications.

Unfortunately, most building operators do not know the actual ventilation rates being supplied by their air handling systems. Most simply set the minimum damper position based on what looks right. At best, they may position the outside air damper at a set

Figure 8.1
Recommended Ventilation Rates

Area	Ventilation Rate (cfm/person)
Ballrooms	25
Bars/cocktail lounges	30
Beauty shops	25
Classrooms	15
Dining rooms	20
Hospital patient rooms	25
Hotel conference rooms	20
Hotel guest rooms	30
Libraries	15
Office conference rooms	20
Office spaces	20
Operating rooms	30
Reception areas	15
Retail stores	(0.02-0.30 cfm/sq ft)
Smoking areas	60
Spectator areas	15
Theaters	15

Note: Reprinted by permission from ASHRAE Standard 62-1989, Standards for Natural and Mechanical Ventilation.

location, assuming a certain percentage of ventilation air. Changes in damper position are made only if there are frequent complaints from the building occupants or if the capacity of the system cannot adequately condition the ventilation air.

There are several problems with both of these approaches to setting the ventilation rate. Dampers are difficult to set up and position with any accuracy. A damper that appears to be open 5 or 10 percent may actually be allowing ventilation rates as high as 50 percent. Even dampers that are fully closed can allow ventilation rates as high as 30 or 35 percent as the result of leakage. With no direct feedback from the operation of the system, other than occupant complaints, there is no way that they can determine the actual ventilation rates. Ventilation rates remain constant, regardless of occupancy and actual need. The result is that most systems are usually overventilated.

Determining the actual ventilation rate is a difficult process involving the use of expensive monitoring equipment. An alternative is to estimate the percentage of outdoor air being introduced into the facility by measuring three temperatures within the system: the return air temperature, the mixed air temperature, and the outside air temperature. Knowing these three temperatures, the percentage of outdoor air being brought into the facility may be calculated from the following:

Formula 8.1

$$\text{Percent outside air} = \frac{(\text{return air temp}) - (\text{mixed air temp})}{(\text{return air temp}) - (\text{outside air temp})}$$

In order for the estimate to be accurate, all temperature measurements must be made with an accurate thermometer. Even minor variations in measured temperatures will result in significant errors in the calculated percentage of outside air. Additionally, temperature measurements must be made at locations where the temperature is uniform across the air flow. While this condition is easily met for return air and outside air temperature measurements, it is more difficult for mixed air temperatures. Mixed air temperature measurements should be taken at a point sufficiently downstream from the dampers to permit complete mixing of the outside and return air flows.

If the actual flow rate of the air handler is known or can be measured accurately, the following formula can be used to determine the ventilation rate:

Formula 8.2

$$\text{Ventilation rate} = \frac{(\text{Total air flow}) \times (\text{Percent outside air})}{(\text{Occupancy})}$$

where
    Total air flow   =   air handler flow rate in cfm
    Percent outside air =   value calculated by formula 8.1
    Occupancy   =   the number of people in the area served by the air handler

Heating and Cooling Coils

The coils used in building air handling systems are low-maintenance components. With a minimum level of effort, coils can last the lifetime of the air handler. However, if they are ignored, they will leak, perform inefficiently, and eventually fail. Replacing coils is an expensive and time-consuming process. It takes time to locate a suitable replacement for a particular air handler. Replacement is not an easy task as the coil sections of most air handlers are not readily accessible. The result is lengthy system downtime in the event of a coil failure.

One of the most important maintenance tasks for heating and cooling coils is keeping them clean. Dirt from the air flow tends to accumulate on the surface of coils. Any accumulation of dirt decreases the heat transfer efficiency of the coil and decreases the air flow in the system. Dirt also collects and holds corrosive materials, promoting the corrosion of the coil surface. A program of regular inspections and cleaning will help ensure that the coil surfaces remain clean and free of dirt and other debris. ***Rule of Thumb:*** At least once a month, heating and cooling coils should be inspected for cleanliness. At least once a year, coils should be cleaned. During the annual cleaning, the cooling coil drain pan should be closely inspected for leaks and rusting.

If periodic inspections show frequent loading of coil surfaces with dirt and debris, it will be necessary to identify the source of the material. Common causes for coil dirt accumulations are the use of the wrong type of air filter, poor filter maintenance, and leaks in the ductwork between the filters and the coils. If the coils are to be properly maintained, it will be necessary to first identify the source of the dirt and eliminate as much as possible from the airstream.

It is also necessary to keep the internal surfaces of the coils clean to maintain their operating efficiency. Direct expansion coils (dx) will remain clean as long as the refrigerant flowing through them is not contaminated. To maintain the cleanliness of coils fed by steam, chiller water, or hot water, an effective water treatment program must be in place. Periodic testing of water samples will determine the effectiveness of the water treatment program in protecting the internal surfaces of the coils.

Filters

In order to understand the maintenance requirements of filters, it is first necessary to understand the role that filters play in air handling systems. For years, the primary purpose of installing filters in the air stream was to protect components of the HVAC system from fouling. Filters were designed to remove dust and dirt from the airstream before they could collect on heating and cooling coils or accumulate on duct surfaces.

Today the role played by air filters is much broader. Filters are expected to protect the finishes of conditioned spaces by removing dust and dirt before they can stain ceilings, walls, and fabrics. They are used to protect the building occupants from exposure to excessive levels of a wide range of contaminants, including respirable particulate matter, pathogens, allergens, odors, and irritating chemicals. Filters are also used to protect particularly sensitive processes from unacceptable levels of contaminants.

In order to fulfill the wide range of roles they are expected to play, a number of different types of filter have been developed. Proper filter performance depends to a great extent on the selection of the proper type for the particular application. Proper filter maintenance depends on the type of filter installed.

Panel filters are the lowest in first and maintenance costs. Units are available in both disposable and permanent models. One of their major disadvantages is low operating efficiency. Panel filters simply do not capture and hold a very high percentage of the particulate matter passing through them. Panel filters, as they load up with dust and dirt, can create a relatively high pressure drop within the system, restricting air flow. In constant volume systems, a heavily loaded panel filter decreases the quantity of air supplied to the conditioned spaces. In variable volume systems, the high pressure drop caused by a loaded panel filter causes the fan to operate at a higher setting in order to meet the heating and cooling loads in the conditioned spaces, increasing fan energy requirements.

Proper maintenance of panel filters requires periodic inspections to ensure that all filter media are properly in place and secure. Inspect for leaks between adjacent panels and between panels and their mounting frames. While many organizations establish a schedule for changing panel filter, such as two or three times a year, it is more efficient to schedule filter changes based on their loading. Panels should be changed when the pressure drop across them reaches a predetermined value as established by the filter manufacturer for that application. More frequent filter changes increase operating costs. Less frequent filter changes result in excessive pressure drop across the filter and reduced air flow in the air handling system.

Disposable roll filters are higher in first costs and maintenance costs than panel filters, but offer improved operating efficiency and a relatively constant pressure drop. Clean filter media are supplied from a roll on one side of the air handler across the air stream. Dirty media are wound on a roll on the other side of the air handler. That rate at which media are fed across the air stream is set according to the level of dirt present in the air. Often it is regulated by the pressure drop measured across the filter, increasing the rate at which media are fed as the pressure drop increases. The primary drawback of disposable roll filters is the need to access both sides of the air handler in order to change filter media.

The rate at which media are fed by roll filters depends on the rate of loading. The most effective means of controlling the feed rate is by measuring the pressure drop across the filter. If an automatic feed system is used, the measured pressure drop across the filter will establish the filter media feed rate. If the feed rate is manually set, monitor the pressure drop across the filter weekly, changing the feed rate until the desired pressure drop is achieved.

Extended surface filters, also know as bag filters, use filter media shaped into elongated bags. The use of the bag design provides for a large increase in filter surface area exposed to the air flow without increasing the cross section area of the air handler.

Extended surface filters are higher in efficiency than either panel or roll filters, but have much higher first and maintenance costs. The use of extended surfaces helps to lower the pressure drop across the filters even as they become loaded with dirt. To lengthen the useful life of extended bag filters, inexpensive panel filters are often installed upstream to prefilter larger contaminants from the air flow.

Maintenance of extended surface filters requires monitoring of the pressure drop across both the bag filters and the panel prefilters. Change the filters when the measured pressure drop approaches the maximum values recommended by the system manufacturer. It will be necessary to change the panel prefilters more frequently than the bag filter as they tend to load up with the larger particles.

Electronic filters offer the benefits of high efficiency, low pressure drop, and low maintenance costs. Their primary drawback is high first cost. Electronic filters use electrostatic charges to attract particles to collecting plates. As the air flow enters the filter, a high-voltage ionizing wire imparts a charge to particles in the air flow. The air then passes through a series of charged and grounded plates that attract the charged particles. An adhesive coating holds the particles to the plates.

The only maintenance electronic filters require is the periodic removal, cleaning, and recoating of the charged plates. The frequency at which the plates must be cleaned depends on the cleanliness of the air flow. Periodic inspection of the filter will help determine the optimum time for cleaning.

Controls

One of the most important components for efficient and proper operation of an air handler is the control system, yet controls are routinely ignored when it comes to performing building maintenance. No matter how much money is invested in efficient boilers, chillers, energy management systems, and other energy-efficient equipment, overall energy use will not be minimized if the control system is not properly maintained. Studies of building energy use have shown that more money is wasted through poorly operating controls than through any other single cause. In contrast, properly maintained controls will reduce energy use, decrease the number of complaints from the building occupants, prolong the useful life of the HVAC equipment, and reduce the amount of time spent on breakdown maintenance.

In pneumatically operated systems, the key component for efficient and effective operation of the entire control system is the compressor. It is essential that the air compressor supply air that is free of oil, water, and other contaminants, and supply it at a constant pressure. Even small quantities of contaminants can interfere with the proper operation of controllers and actuators.

In addition to maintenance of the building's control air compressor, there are two major maintenance activities that involve the control system for the air handler: verification of control operation and calibration. Verification involves testing the operation of controlled devices, such as valves and damper motors, and all system safety devices to ensure that they operate as intended. Too often devices fail to respond to control system commands as the result of device failures or something as simple as being bypassed by maintenance personnel in an attempt to correct a complaint. If the system is to operate efficiently, devices must respond to command from the control system. Annual verification of device operation helps ensure that they will.

The second annual maintenance activity, calibration, is designed to ensure that when devices respond to commands from the control system, they respond exactly as intended. Control devices can easily drift out of setting as the result of normal wear, or as the result of contamination by dirt. Out of calibration devices result in the system supplying more heating, cooling, or ventilation air than is actually required, wasting energy and increasing the number of trouble calls from occupants.

Rule of Thumb: The best time to perform verification testing and device calibration is during the late spring or early fall when the system is not operating at full capacity and the control system is modulating the supply of heating or cooling energy. At least once a year, all control devices and actions should be verified and calibrated.

Dampers

The dampers in an air handler are designed to provide a specific minimum ventilation rate in order to maintain a minimum acceptable level of indoor air quality. Too low a rate and air quality suffers. Too high a rate, and energy use in the system increases. Unfortunately, few systems are properly calibrated when new. With time and use, they only become worse, typically introducing more ventilation than actually is required. In some cases, maintenance personnel have wired dampers to a set position in an attempt to "fix" maintenance problems. The result is usually excessive energy use.

Rule of Thumb: To ensure proper operation, all system dampers should be inspected at least every three months. All dampers must be able to rotate freely without binding. All damper seals must be intact and must properly seal when the damper is in the fully closed position. Every three months, inspect dampers for proper operation throughout their entire range. Clean and lubricate all moving parts.

Ductwork

Although the ductwork is a passive component in an air handler system, it too requires a program of ongoing maintenance inspections. There are two primary concerns with the ductwork: leaks and damaged or missing insulation.

Any air losses from ductwork reduce the volume of air that is provided to the conditioned space, resulting in decreased air quality and increased energy use. There are a number of common areas for leaks in both metal and fiberglass ducts, including joints that have separated, access and inspection plates that were removed and never reinstalled, construction and renovation damage, and holes caused by hangers tearing loose. At least once a year, the ductwork connected to the air handler should be closely inspected for leaks. It is easier to survey the system for leaks during the cooling season when leaks would produce abnormally cool areas, such as above ceiling tile.

The condition of the insulation is important to both the performance and the life of the ductwork. During the cooling season, areas with damaged or missing insulation tend to sweat. The moisture that is formed promotes corrosion of metal ductwork and deterioration of fiberglass ducts. ***Rule of Thumb:*** During the annual inspection for leaks, the ductwork should also be inspected for missing and damaged insulation. While damage to exterior duct insulation is easily detected, many systems have been installed with interior duct insulation that cannot be seen during a routine inspection. For systems with interior duct insulation, inspect for signs of corrosion and sweating on the surface of the duct. Corrosion and sweating are signs that the interior insulation is no longer functioning properly.

HOW TO ESTABLISH THE MAINTENANCE PROGRAM FOR AIR HANDLERS

The first step in developing a maintenance program for air handlers is to identify the location of all systems in the facility. For each system, use Figure 8.2 to record basic data on the air handler installation. Depending on the size of the facility, the number of air handlers installed, and the past level of maintenance performed, it may not be possible to implement the program for all air handlers at the same time. Start-up costs and the labor required to implement a new program may be too high for many organizations. Therefore, it will be necessary to phase the program in over several months or even years.

Using the list of installed air handlers, determine the priority for implementing the maintenance program for each air handler. Factors that must be considered when determining the priorities for individual air handler include energy use of the system, the current condition of the air handler, the critical nature of the operation served by the air handler, and the level of IAQ problems currently being experienced in the area served by the air handler.

Once the systems have been inventoried and priorities have been established, identify the necessary tasks to be performed. Figure 8.3 lists recommended maintenance activities to be performed monthly for air handlers operating under normal service conditions. Figure 8.4 lists recommended maintenance activities to be performed annually. Use these figures as a starting point in developing the maintenance program. It may be necessary to modify the tasks to be performed and the frequency with which they are performed for specific systems based on the particular application.

In addition to the tasks identified in the figures, it will be necessary to set up a program for monitoring the operation of the air handling systems. Record all maintenance activities performed, including those performed on a scheduled basis and those performed as the result of a breakdown or component failure. All complaints received regarding the operation of a particular system should be recorded and tracked. Track energy use, particularly if there is sufficient metering to track to energy requirements of individual buildings or systems. Periodic review of these maintenance records will help in identifying developing problems as well as systems that are in need of a major overhaul or total replacement.

Figure 8.2
Air Handler Inventory Data

1. Building/facility: _____

2. Room #/location: _____   3. Air handler #: _____

4. Type of system: _____

5. Application: ☐ entire building ☐ partial building ☐ special application

6. Function: ☐ heating only ☐ cooling only ☐ heating and cooling

7. Airflow rate (cfm): _____

8. Supply fan horsepower: _____

9. Return fan horsepower: _____

10. Type of controls: ☐ pneumatic ☐ electronic ☐ direct digital control

11. Type of filter: ☐ panel ☐ roll ☐ bag ☐ electronic

12. Manufacturer: _____

13. Model: _____

14. Date installed: _____

15. Comments:_____

_____

_____

_____

_____

_____

_____

Figure 8.3
Monthly Air Handler Maintenance Activities

1. Building/facility: _____

2. Room #/location: _____ 3. Air handler #: _____

4. Dates: _____

	Month					
Activity	**1**	**2**	**3**	**4**	**5**	**6**
Lube fan bearings						
Lube motor bearings						
Check fan for noise and vibrations						
Inspect drive sheaves for wear						
Inspect coils for cleanliness						
Inspect coils for leaks						
Inspect drain pan for ponding						
Inspect drain pan for debris						
Check drain line for obstructions						
Measure pressure drop across filters						
Inspect filter media						
Inspect fit of filters						
Inspect and cycle dampers						
Inspect damper seals						
Clean and lube damper actuators						
Check outside air intakes for obstructions						
Clean temperature and humidity sensors						
Check control lines for oil and water						
Test operation of freeze-stat						
Check access doors for tightness						

Figure 8.4
Annual Air Handler Maintenace Activities

1. Building/facility: _____

2. Room #/location: _____ 3. Air handler #: _____

4. Dates: _____

Activity	Completed	Comments
Inspect fan drive for wear	☐	_____
Check fan drive alignment	☐	_____
Inspect fan for rust	☐	_____
Inspect fan blades for damage	☐	_____
Clean fan wheel	☐	_____
Measure pressure increase across fan	☐	_____
Measure fan speed	☐	_____
Measure motor current	☐	_____
Inspect motor mounts	☐	_____
Clean coils	☐	_____
Inspect coils for leaks and cracks	☐	_____
Inspect coil frames for corrosion	☐	_____
Clean drain pan	☐	_____
Inspect drain pan for rust	☐	_____
Inspect filter frame	☐	_____
Verify controls operation	☐	_____
Calibrate controls	☐	_____
Check duct interior for water damage near coils	☐	_____
Inspect duct work for leaks	☐	_____
Inspect duct insulation	☐	_____
Check fire damper positions	☐	_____
Inspect vibration isolators	☐	_____
Clean electrical contacts	☐	_____

It will take time for the results of the maintenance program to take effect, particularly if maintenance requirements have been largely ignored in the past. Initially there will be an increase in maintenance requirements, particularly as maintenance personnel play catch-up with ongoing but previously undetected maintenance problems. With time, though, the program's effects will start to be felt in terms of reduced breakdown maintenance requirements, reduced energy use, and reduced occupant complaints.

FIVE OPERATING PRACTICES FOR AIR HANDLING SYSTEMS TO ENHANCE ENERGY CONSERVATION

How building air handler systems are operated to a great extent determines their seasonal energy efficiency. By implementing specific practices, you can reduce the annual energy use of the systems by as much as 25 percent without negatively impacting the operations taking place within the areas served by the air handler, or by causing discomfort to the building occupants. Most of these practices are the result of simply paying attention to detail in the day-to-day operation and management of the system, such as ensuring that thermostats are properly set and that maintenance activities are properly scheduled and completed. Others involve implementing practices specifically designed to reduce energy use. Both require constant attention and periodic review to ensure that they are being followed satisfactorily.

Although there are a wide range of operating practices that can be implemented to reduce seasonal energy requirements, five of the most common and successful include scheduled operation, optimum start–stop timing, day–night setback, economizer cycle, and recirculation control. While not all of these operational practices are suitable for all air handler applications, many facilities can benefit from their use. Before implementing any of the practices, determine their impact on the operations taking place in the conditioned space.

Scheduled Operation

Nearly all facilities have set hours of operation; few are required to operate twenty-four hours per day, seven days per week. For example, most office facilities are occupied no more than ten or eleven hours per day, five days per week. That means that more than 50 percent of the time annually, the facility is unoccupied. Matching the hours that the air handler is operated to the hours that the facility is occupied will save heating and cooling energy as well as fan energy.

Under scheduled operation, systems are started sufficiently prior to occupancy to allow the conditioned spaces to be warmed up or cooled down. Typical warm-up and cool-down periods range between one-half and one hour, depending on the capacity of the system and the heating or cooling load imposed by the occupied space.

The savings produced by scheduling the operation of the air handling units to the occupancy of the facility will vary with the climate of the facility and the number of hours that the system can be safely shut down. Typical annual savings from scheduled operation range between 10 and 15 percent of the energy used by the system.

Before the operation of air handlers is scheduled, maintenance personnel must ensure that the system's dampers are operating and are properly sealing outside air from the system. A stuck or leaking outdoor air damper can allow sufficient cold outdoor air to enter the system and freeze the unit's coils. Additionally, long shutdown periods may cause space temperatures to fall sufficiently during the heating season to cause damage in the occupied spaces. Similarly, long shutdown periods during the cooling season, particularly in humid climates, may cause damage to interior finishes from excessive humidity.

Optimum Start–Stop Timing

This operating practice is a refinement of scheduled operation. Under scheduled operation, the air handlers are started at a pre-established time in order to sufficiently warm or cool a facility by the time it is occupied. The scheduled start time is typically established far enough in advance of building occupancy so that the proper temperatures can be achieved for even the hottest or coldest days.

Optimum start–stop operation modifies the system start time based on the actual conditions of the day. Most systems monitor outside air temperatures, outside air humidity (primarily for cooling), and temperatures within the conditioned space. The system then calculates the amount of time required to bring the occupied spaces up to or down to the desired temperature by the time the facility is scheduled to be occupied. For example, a facility may require bringing the air handlers on line thirty minutes before occupancy when outside temperatures are at 20 degrees, but may require only ten minutes lead time when the outside temperature is between 50 and 60 degrees.

The savings produced by optimum start–stop operation typically range between 1 and 2 percent of the annual energy use of the system in addition to the savings produced by scheduled operation. Most of the savings are in the form of reduced fan energy and reduced energy required to heat or cool ventilation air.

Day–Night Setback

The day–night setback operating practice is an alternative to the practice of completely shutting down air handlers when the facility is unoccupied. Some facilities, particularly those having processes or equipment that are sensitive to wide swings in temperature, may not be suitable for unoccupied hours shutdown of the air handlers. Day–night setback, in contrast, simply raises the temperature of the air supplied to the conditioned space during the cooling season, and lowers the temperature during the heating season. The amount of temperature change is determined by what the facility can tolerate and the length of time that the facility is to be unoccupied.

By controlling the temperature during the unoccupied period, problems with low or high space temperatures or high space humidity levels are avoided. The savings produced, however, are less than what is achieved through unoccupied hours shutdown as the air handler is operating and using energy. Additional energy savings can be achieved by closing the outdoor air dampers while the facility is unoccupied. Actual energy savings vary with the amount of temperature setback implemented and the number of hours that the system is operated in day–night setback. For most applications, you can expect to achieve annual savings of 5 to 10 percent for day–night setback use.

Economizer Cycle

One of the more effective operating practices for reducing the cost of building air conditioning is the use of an economizer cycle. Economizer cycle operation is a way of achieving free cooling by using outside air. When the total energy content of the outside air (temperature plus humidity) is less than the total energy content of the building's return air, the system increases the amount of outside air introduced into the building, reducing the cooling load on the system. If outside air temperatures are low enough, the system can provide free cooling by simply mixing the outside air with the building return air to obtain the desired supply air temperature.

Economizer cycle operation is particularly effective during late spring and early fall when air conditioning is needed to maintain building temperatures while outside air temperatures are fairly moderate. The system is also effective during early morning hours before the outside air temperature have risen above the desired building air temperature. During these periods, economizer cycle operation is used to precool the facility for free before it is occupied. Depending on the climate, several hours of free cooling may be obtained on many summer days. The typical annual savings that can be expected to produced by use of economizer cycle operation is around 10 percent of the annual cooling cost for the system.

The most serious drawback of the system is the lack of humidity control during operation. Building air conditioning systems rely on mechanical cooling to dehumidify the air supplied by the air handler. Under economizer cycle operation, there is no method of controlling the humidity of the air supplied to the conditioned space. Monitoring space humidity conditions during economizer cycle operation will provide feedback that can be used to determine when it is best to switch to conventional mechanical cooling.

SUMMARY

Building air handling systems are critical elements in maintaining safe environments for the building occupants to work in. Their proper maintenance will help ensure that indoor air quality is maintained at an acceptable level without excessively high levels of energy use. By properly maintaining air handling systems you can achieve energy savings typically ranging from 10 to 25 percent without compromising indoor air quality. Additional savings can be achieved through implementation of energy conserving operating practices. Start-up costs may be high, particularly if maintenance has been ignored in the past and there are a large number of maintenance deficiencies in need of correction. As with other maintenance programs, paybacks for the investment in maintenance are measured in months not years.

Chapter 9

Service Hot Water Systems

This chapter presents guidelines on maintaining service hot water systems. The information presented may readily be applied to the types of service hot water systems typically found in commercial, industrial, and institutional facilities. It should be noted that the maintenance activities identified in this chapter and their recommended frequencies represent the minimum recommended level of maintenance to be performed on these systems. Systems supporting critical operations will require additional and more frequent attention. If additional information is required in identifying the recommended level of maintenance for a particular piece of equipment in a system, it is recommended that the equipment manufacturer be contacted.

For the purposes of this chapter, the components of a service hot water system are defined to include the system's water heater or heat exchanger, the control system, the storage tank, and the distribution system. While the individual maintenance requirements will be included in the discussion of the operation of each component in the system, the recommended system maintenance activities will be presented later in this chapter.

SERVICE HOT WATER SYSTEM ENERGY MAINTENANCE CONSIDERATIONS

Service hot water systems, or domestic hot water systems, as they are sometimes called, are a frequently overlooked component in building maintenance and energy management. Ignoring service hot water systems is often justified on the basis of the relatively low energy use rate in comparison to other energy using systems within the facility. In most facilities, service hot water energy use is less than 10 percent of the total energy use of the facility. But while service hot water energy use may be small in comparison to the energy used by other systems, it is still significant. Equally important, major reductions in energy use can be achieved easily and painlessly through a well planned and executed maintenance program. Energy use in well maintained systems is often only 50 percent of that in systems that have been ignored for even a few years.

Cutting energy losses in service hot water systems also saves through reduced water use. Two of the consequences of poor maintenance practices include leaks and higher than necessary hot water use. By implementing a maintenance program for service hot water systems, leaks will be identified and water use levels will be matched to the need. Often, the savings achieved through reduced water use will exceed the energy savings.

Another factor that contributes to ignoring the maintenance of service hot water systems is the function that the systems perform. As long as a system is working and supplying hot water to where it is needed within the facility, nobody cares. Water temperatures can vary by 20 degrees or more and it will not impact most operations or trigger a rash of complaints. About the only concern on the part of the occupants is if the temperature gets high enough to pose a threat from scalding, or if the temperature gets too low for certain applications, such as laundry or dishwashing.

Even though the energy used by service hot water systems is small in comparison to the energy used by other systems in the facility, it is important that they be properly maintained. Without regularly scheduled maintenance, water heaters and heat exchangers become scaled and plugged. Piping and fittings develop leaks. Insulation loses its effectiveness from deterioration and physical damage. Circulation pumps leak and fail. Faucets and valves wear and fail to fully close. The result is increased use of service hot water, increased energy costs, and decreased system reliability.

Increased energy and water costs are not the only consequences of neglected maintenance. Without proper maintenance, the performance and operation of system components declines and component life is shortened. On an even more serious level, neglected maintenance can result in conditions that are hazardous to the facility occupants.

SERVICE HOT WATER REQUIREMENTS

Nearly all buildings make use of service hot water, even if it is used only for hand washing. Some facilities, such as hospitals, hotels, dormitories, and those with food service or laundry facilities, have extensive needs for service hot water. Service hot water requirements in industrial facilities may result in energy costs that approach or exceed those for heating or air conditioning. Figure 9.1 lists the average per day service hot water requirements for various types of facilities. The values presented in the figure are for typical facilities with service hot water systems that are in very good condition. None of the values listed takes into consideration increased use of service hot water resulting from improperly operating equipment or leaks.

In addition to variations in the demand for service hot water, there are variations in the temperature at which the water must be supplied. Figure 9.2 lists typical service hot water requirements for various end uses. It should be noted that the values listed in the figure are the temperature of the water as measured at the point of use. In most system designs, it may be necessary to operate the water heater or the system's heat exchanger at a slightly higher temperature in order to obtain these outlet temperatures as a result of losses in the piping system.

Figure 9.1
Average Service Hot Water Requirements for
Various Building Types

Type of Building	Daily Average
Men's dormitories Women's dormitories	13.1 gal/student 12.3 gal/student

Motels
    20 units or less
    60 units
    100 units
	
	20.0 gal/unit
	14.0 gal/unit
	10.0 gal/unit

Nursing homes	18.4 gal/bed
Office buildings	1.0 gal/person

Food service
    full meal restauraunts & cafeterias
    drive-ins, grills, luncheonettes, snack shops
	
	2.4 gal/meal
	0.7 gal/meal

Apartment houses
    20 apartments or less
    50 apartments
    75 apartments
    100 apartments
    100 or more apartments
	
	42.0 gal/apartment
	40.0 gal/apartment
	38.0 gal/apartment
	37.0 gal/apartment
	35.0 gal/apartment

Schools
    Elementary
    Junior and senior high
	
	0.6 gal/student
	1.8 gal/student

Source: Reprinted by permission from *1995 ASHRAE Handbook, HVAC Applications.*

Figure 9.2
Recommended Service Hot Water Temperatures

Use	Temperature (^{o}F)
Lavatory	105
Showers and tubs	110
Commercial and institutional laundry	<180
Surgical scrubbing	110
Commercial dishwashers	150-180

Source: Reprinted by permission from *1995 ASHRAE Handbook, HVAC Applications.*

Surveys of facilities have shown that actual service hot water system temperatures are well above the recommended supply temperatures. The average temperature of systems operating in office facilities, where the major use of hot water was for hand washing, was approximately 130 degrees Fahrenheit. In other facilities, such as hotels and motels, the average temperature of systems feeding the guest rooms was 140 degrees Fahrenheit. In facilities that used service hot water in sanitizing operations, such as laundry or dish washing, the supply temperature was between 180 and 190 degrees Fahrenheit.

The temperature at which service hot water is supplied is particularly important to the energy efficiency of the entire system. To raise the temperature of a gallon of water 100 degrees Fahrenheit requires 830 Btu. At an average operating efficiency of 85 percent, a water heater would require approximately 976 Btu of energy input for each gallon of hot water produced. If the hot water supply temperature is reduced by ten degrees, the energy requirement for the water heater would be lowered to 878 Btu per gallon, a 10 percent savings. Similarly, a 20 degree decrease in hot water supply temperature results in a decrease in the energy requirement for the water heater by 20 percent to 780 Btu per gallon. Therefore you should pay particular attention to the temperature requirements for service hot water and the temperature at which systems are operating.

TWO SERVICE HOT WATER SYSTEMS

There is a large variety in the types of service hot water systems that are used in commercial, institutional, and industrial facilities. Systems are typically fueled by natural gas, steam, electricity, or oil. Systems can be centrally located within the facility, or individual water heaters can be installed at each point of use. The majority of the systems fall into two categories—storage systems and tankless systems.

Storage Systems

Storage systems consist of an insulated tank, a heating element or burner, and a distribution system. Depending on the size of the system, the heating element may be part of the storage tank, such as with the typical residential water heater, or it may be located external to the tank with the heating of the water taking place in a heat exchanger. Also depending on the size of the system, two different distribution designs are used—noncirculating and circulating.

Noncirculating systems, similar to residential hot water systems, rely on the pressure in the water system to distribute the service hot water to the various points of use. The systems are low in first costs, have no moving parts, and are easy to maintain. They are well suited to smaller facilities where the distance between the hot water storage tank and the points of use of the hot water are short, and there is little pressure loss in the distribution system.

Noncirculating systems are not well suited for larger facilities, where long piping runs are required between the storage tank and the points of use. As the distance grows, the water tends to cool in the piping during periods when it is not being used. When a user needs hot water, the water must then be run for a period of time to allow the water to reach the desired temperature, wasting both water and the energy used to heat it. Also, noncirculating systems are not well suited for facilities much higher than three or four stories due to the pressure drop in these installations.

Circulating systems use a pump to distribute service hot water from the storage tank to the points of use within the facility. They are higher in first costs than noncirculating systems due to their need for a return line and a pump, require additional energy to pump the water through the system, and are only slightly more expensive to maintain. Circulating systems are typically used in medium to large or multi-story facilities to compensate for the temperature and pressure losses found in noncirculating systems.

Since the water is pumped through the system, temperatures remain fairly constant with few losses between the storage tank and the points of use. The result is that when service hot water is turned on at any point in the facility, the user does not have to wait for the water temperature to rise to the desired point, cutting water use.

Tankless Systems

Tankless systems, also known as instantaneous, consist of a heat exchanger, circulation pump, and distribution system. Water is circulated between the heat exchanger and the points of use in the system. To maintain a constant supply temperature, mixing valves are used in the hot water supply. A commonly used modification of the tankless system is the semi-instantaneous system, which makes use of limited amount of storage in order to offset momentary surges in demand for hot water.

Tankless systems are best suited for applications where there is a relatively constant demand for hot water. Without a storage tank, system temperatures can vary widely if there are surges in demand.

ELEMENTS OF THE SERVICE HOT WATER SYSTEM ⸻

Although there are a number of different designs and system configurations, most systems have the same basic components. For the service hot water system to operate as efficiently as possible, all of these components must be in good working order. Maintenance of these components will not vary significantly from one system to another.

The service hot water systems maintenance program must focus on several key elements in order to be effective. These elements include the water heater, the storage tank, the distribution system, and the end uses of the hot water. The maintenance requirements for each of these components will be discussed in the following sections. Suggested maintenance activities for all components will be provided in a later section of this chapter.

Water Heaters

The majority of water heaters used in facilities today are fueled by natural gas, oil, electricity, or steam. The overall efficiency of the entire service hot water system depends to a great extent on the operating efficiency of the water heater. A well-maintained water heater can be expected to operate with an average efficiency ranging between 80 and 85 percent for natural gas, oil, and steam fueled units, and nearly 100 percent for units fueled by electricity. Units that are not well maintained operate at much lower efficiencies, sometimes as low as 50 percent.

One of the most important factors in keeping water heaters operating at their peak efficiency is cleanliness. City water supplied to the water heater contains mineral salts and other contaminants. As the water is heated, these contaminants precipitate from the water, forming scale on heating surfaces, reducing their heat transfer efficiency. If the scale is allowed to build unchecked, eventually it can completely clog heating coils.

There are several maintenance activities that can be implemented to reduce the impact that scale has on water heater efficiency. First, steps should be taken to reduce the rate at which scale forms. The rate of scale formation is determined in part by the temperature of the water produced by the heater and the hardness of the water supplied to the heater. Keeping the temperature of the water below 140 degrees Fahrenheit significantly slows the rate at which scale forms, so unless there is a specific need for higher temperatures, water heaters should be operated at 140 degrees Fahrenheit or less.

A second way to reduce the rate of scale formation is through the use of water softeners. Treating cold water with softeners before it enters the water heater reduces the level of salts that are dissolved, lowering the hardness of the water and the rate at which scale is formed on heat transfer surfaces.

In addition to taking steps to reduce the rate at which scale is formed, the water heater should be examined for scale buildup and cleaned if necessary. Even a layer of scale as thin as one-sixteenth of an inch will reduce the efficiency of the water heater by approximately 6 percent. *Rule of Thumb:* At least once a year, water heaters should be inspected for scale and cleaned as required.

The operating efficiency of natural gas and oil fueled water heaters also depends on the efficiency of the burner. While the burners are simple and reliable, they occasionally need adjustment. *Rule of Thumb:* Temperature controls should be inspected monthly. At least once a year, burners should be inspected, cleaned, and adjusted.

Steam-fed water heaters must also be maintained. *Rule of Thumb:* Once a month, the temperature control system should be inspected and adjusted as required in order to maintain proper supply temperatures within the system.

Storage Tanks

Storage tanks are passive devices in the service hot water system but still require maintenance if they are to perform efficiently throughout their rated life. The most common types used in facilities are steel, glass-lined steel, and fiberglass. Regardless of the type installed, a well-maintained storage tank can be expected to operate efficiently for twenty years or more while a poorly maintained tank will have excessive heat loss, reduced capacity, and increased risk of failure from internal corrosion.

One of the most important maintenance activities regarding storage tanks is the removal of sludge from the bottom of the tank. Sludge is formed as solids precipitate out of solution from the heated water and settle on the bottom of the tank. If allowed to remain in the tank, the sludge will accumulate, reducing the capacity of the tank and corroding its interior surface. ***Rule of Thumb:*** At least once a year, the tank should be flushed to remove all sediment.

The rate of corrosion of steel tanks varies with the temperature of the water stored in the tank. Higher temperatures result in higher rates of corrosion. When water temperatures exceed 140 degrees Fahrenheit, two factors increase the rate of corrosion even further. At temperatures above 140 degrees Fahrenheit, the conductivity of the water increases significantly, increasing the rate at which galvanic corrosion takes place. Also, at temperatures above 140 degrees Fahrenheit carbonates dissolved in the water break down, creating carbonic acid, which readily attacks metal surfaces.

The energy efficiency of the tank is determined by the effectiveness of its insulation. Assuming that the tank was properly insulated when it was installed, insulation can reduce heat losses from the tank by more than 90 percent. But insulation is easily damaged by leaks and maintenance activities performed in the area of the tank. ***Rule of Thumb:*** Once a year, the tank insulation should be inspected and repaired as necessary.

The Distribution System

The distribution portion of a service hot water system is a low-maintenance item. Typically constructed from copper, galvanized, or plastic piping, the piping will last between 30 and 50 years under normal use. Actual service life is determined more by the hardness of the water than by maintenance performed on the piping. While the piping itself needs little or no maintenance, there are two areas requiring regular maintenance in order to maintain energy efficiency: inspection of the system for leaks and inspection of the pipe insulation.

Most leaks in service hot water distribution systems occur not in the pipe itself, but rather in valves and other fittings. It is not uncommon to find a large percentage of the valves installed in the system leaking at their packing gland. While these leaks may appear to be insignificant, the volume of water and quantity of energy lost over the course of a year is not. For example, a small leak that drips about once a second looses approximately 2,500 gallons of water a year. A leak at the rate of a small stream of water from a valve or fitting can lose as much as 25,000 gallons in a year. All leaks must be identified and repaired as quickly as possible if energy and water losses from the distribution system are to be minimized.

Keeping the pipe insulation in good condition also is important to conserving energy in the distribution system. In poorly insulated systems, particularly circulating systems in applications where the demand for hot water is low, distribution losses frequently exceed actual use from the system. The actual heat loss from the distribution systems depends on the length of the piping, the condition of the insulation, the temperature of the water being circulated in the piping, and the ambient air temperature where the piping is located. For example, a one-and-one-half-inch pipe with no insulation, carrying service hot water at a temperature of 140 degrees Fahrenheit through an area where the ambient temperature is 70 degrees, loses heat at the rate of 53 Btu per hour per foot of pipe. The rate of heat loss from the same pipe when covered with one-half inch of glass fiber insulation is reduced to approximately 25 Btu per hour per. Thicker insulation reduces it even further.

Rule of Thumb: To maintain the integrity of the distribution system's insulation, the system should be inspected at least once a year for areas of damaged or missing insulation. During the inspection, look for areas where the insulation may be wet. Insulation that is wet may be a sign that there is a leak in the distribution system piping, or a leak in some other piece of equipment. Repair of the leak and replacement of the wet insulation is important as wet insulation loses most of its insulating properties.

End Uses of Service Hot Water

Examining the ways in which hot water is used in the facility is the most important step in reducing energy use of the service hot water system. To perform this task successfully requires a walk-through inspection of the entire facility, noting how hot water is used, how much is required, and at what temperature it is to be supplied. Include all uses of hot water for industrial washing and process cleaning, such as dishes and laundry, as well as for personnel needs. Use Figure 9.3 to identify hot water requirements in the facility.

Once the way in which service hot water is used in the facility has been identified, start with the industrial and cleaning uses first. Examine the temperatures at which they are currently using hot water to determine if lower temperatures can be used. In some cases it may even be possible to do away with hot water simply by changing the cleaning material used.

Check the temperature at which hot water is supplied for personnel needs. If possible, temperatures should be lowered until they equal the values listed in Figure 9.2. If the supply temperature of the entire system must be run higher due to the needs of only one end use, consider operating the system at the lowest temperature possible, and installing an instantaneous water heater to boost the temperature for the one load.

While reducing the operating temperature of the system by matching the supply temperature with the needs of users will save energy, by far the greatest savings will come from lowering the demand for hot water. In most applications, the use of hot water, particularly for personnel needs, far exceeds what is actually required, resulting in wasted water and energy. Even though water conservation efforts may have been implemented at some time in the past, unless particular attention had been paid to maintaining those devices, many may have failed and been removed.

Figure 9.3
Service Hot Water Requirements

Building/facility: _____

Location	Use	Temperature
_____	_____	_____
_____	_____	_____
_____	_____	_____
_____	_____	_____
_____	_____	_____
_____	_____	_____
_____	_____	_____
_____	_____	_____
_____	_____	_____
_____	_____	_____
_____	_____	_____
_____	_____	_____
_____	_____	_____
_____	_____	_____
_____	_____	_____
_____	_____	_____
_____	_____	_____
_____	_____	_____

The first step in reducing the demand for hot water at the point of use is to eliminate all leaks. Leaks at hot water outlets are common. Water valve components are subjected to both wear and stresses. Regular opening and closing of valves wears valve components. Component temperatures are repeatedly cycled between the temperature in the space and the temperature of the hot water system, resulting in stresses in sealing components. Solids suspended in the hot water flow accumulate on valve seats, increasing wear and blocking their ability to fully seal. As a result, it is not uncommon to find that a fairly large percentage of the hot water outlets in a facility are leaking. As discussed earlier, even small leaks over a period of months will waste a large quantity of both water and energy.

Leaks are best detected and corrected through a regularly scheduled inspection program. ***Rule of Thumb:*** At least once a month, all hot water outlets should be inspected for proper operation and closing without leaking. Any faucet or other outlet found leaking should be repaired immediately. Commonly used maintenance parts should be kept in stock.

In addition to eliminating all leaks at hot water outlets, automatic devices should be considered that limit the quantity of water that is allowed to flow from outlets such as faucets. There are two major approaches to limiting this flow: flow regulators and automatic closing devices.

Flow Regulators. Flow regulators limit the rate at which hot water flows from the outlet. Unrestricted faucets, such as those found in washrooms, discharge water at the rate of six to ten gallons per minute—a rate far above what is actually needed for hand washing. Standard shower heads discharge even greater flow rates.

A well-designed regulator will reduce the flow rate of water from a faucet to less than two gallons per minute while maintaining adequate pressure and flow. Similar regulators are available for shower heads that will reduce the flow to three gallons per minute

Automatic Closing Devices. While flow regulators limit the volume of flow from the faucet, automatic closing devices limit the period of time that the water can flow. Two basic types are available: sensing units and timed units. Sensing units automatically detect when an object, such as a hand, is placed under the faucet, starting the flow of water. When the object is removed, the flow stops automatically. Timed units, when activated, allow water to flow for a set period of time before automatically shutting off.

The most energy-efficient option is the use of both devices on as many outlets as possible. The use of both will provide automatic on–off control of water flow while limiting the flow to only the rates that are actually needed. It will take some experimentation to determine the best types of devices to be installed, and the most appropriate flow rates. The combined units typically have a payback of less than one year.

IMPLEMENTING THE ENERGY MAINTENANCE PROGRAM

The maintenance program for service hot water systems is relatively easy to implement. The major requirement is for ongoing inspections consisting of both monthly and annual activities. Unlike many other maintenance activities for energy conservation, it is not necessary to phase in implementation of the service hot water maintenance program. Most service hot water systems are limited in scope and extent. Therefore, program startup costs are minimal. It will be necessary, however, to identify the location and all uses of hot water in the facility.

Once the survey of the uses of hot water has been completed, identify the necessary tasks to be performed. Use Figure 9.4 to identify monthly maintenance and inspection activities. Use Figure 9.5 to identify annual maintenance and inspection activities. The activities listed in these figures are suitable for most building service hot water systems. Additional activities may be required depending on the particular needs of the facility. Maintenance programs must take end use requirements into consideration. The energy conservation impact of the program will vary widely for different facilities. Similarly, the consequences of system breakdown are not the same in office buildings as they are in hospitals.

The effectiveness of the program is best monitored through the use of metering on the fuel supply to the water heater. Reduced operating temperatures, reduced flow rates, and reduced use of service hot water will be identified through reduced demand for energy. Tracking system performance over time will also help in identifying problems as they occur.

SUMMARY

Although building service hot water systems are relatively small users of energy in most facilities, they offer an opportunity to make significant cuts in both energy and water requirements. Typical reductions in both energy and water use that can be achieved by implementing a complete maintenance program range between 25 and 50 percent. Equally important, these savings can be achieved without negatively impacting the processes that require service hot water or the building occupants who use it.

Maintenance programs for service hot water systems are easy to establish and are not labor intensive. Depending on the condition of the existing system and how well it has been maintained in the past, most maintenance programs pay for themselves in less than six months.

Figure 9.4
Monthly Service Hot Water System Maintenance Activities

1. Building/facility: _____

2. Dates: _____

Activity	Month					
	1	2	3	4	5	6
Inspect system for leaks						
Inspect outlet devices for proper shut-off						
Inspect flow regulator operation						
Test operation of automatic closing devices						
Adjust temperature controls						
Test pressure relief valve						
Record energy use						
Record water use						

Figure 9.5
Annual Service Hot Water System Maintenance Activities

1. Building/facility: _____

2. Date: _____

Activity	Completed	Comments
Inspect water heater for scale	☐	_____
Adjust & clean burner	☐	_____
Clean heat exchanger	☐	_____
Drain sludge from tank	☐	_____
Inspect tank lining	☐	_____
Inspect tank insulation	☐	_____
Inspect circulation pump	☐	_____
Lube circulation pump	☐	_____
Inspect pipe insulation	☐	_____
Check water temperature at outlets	☐	_____

Section 3

Electrical System Operation and
Maintenance Practices to
Improve Energy Efficiency

Chapter 10

Lighting Systems

This chapter presents information on maintaining building interior and exterior lighting systems. The procedures in this chapter may be applied to all types of lighting systems found in and around nearly any type of facility, including commercial, institutional, industrial, and multi-family residential. The maintenance activities identified in this chapter and their recommended frequencies represent the suggested level of maintenance effort required to support lighting systems, including those used for security purposes. By following the procedures identified here, you will be able to develop and operate a lighting maintenance program that meets the needs of your particular facilities.

This chapter focuses primarily on indoor lighting systems, as the vast majority of facility lighting energy use is used by indoor systems. However, the same maintenance practices are directly applicable to all outdoor lighting systems.

BUILDING LIGHTING SYSTEM MAINTENANCE CONSIDERATIONS

Building lighting systems account for an estimated 30 to 40 percent of the electrical energy used within facilities. They are second only to building air conditioning systems in electrical energy use. Lighting systems designers spend a lot of time and effort designing systems that will both meet the lighting needs of the occupants and be energy efficient. Building owners spend large amounts of money operating and upgrading existing systems to improve their performance and their efficiency. Lamps and fixtures have been changed. Ballasts have been upgraded to higher-efficiency units. Automated controls have been installed.

In spite of these investments, building lighting systems remain one of the most neglected and poorly maintained systems in buildings. There is no planned approach to lighting system maintenance in most facilities today. Building lighting systems are so poorly maintained that many of the features that building owners invested in to achieve quality and cost-effective lighting systems are negated. In most facilities today, lamps are replaced individually as they fail. Fixtures are cleaned only after occupants raise complaints. Defective lighting fixture components are replaced only after the fixture has stopped operating properly. Diffusers and lenses are seldom replaced, regardless of their condition; many are simply missing.

This reactive approach to building lighting system maintenance is uneconomical and wasteful in terms of both labor and energy use. A typical lighting system that has been maintained by reactive maintenance produces 30 to 40 percent less light than when it was first installed, even though it still uses the same amount of energy. Performing maintenance reactively is five to ten times more costly than a planned maintenance program. As a result, reactive lighting systems maintenance does not allow facility managers to get what they pay for in terms of lighting quality. They paid for the lighting system design and installation, and the existing system will use the same amount of energy regardless of whether it is maintained or not. What really suffers is the quality of lighting provided. Quality that will impact the operations taking place within the facility.

In contrast to reacting to lighting problems, a planned lighting system maintenance program helps to optimize the performance of the lighting system while reducing both maintenance and energy costs. And with lighting levels maintained at a significantly higher level than in facilities where lighting system maintenance is performed reactively, the potential exists to save energy by reducing the light output of the fixtures. Other maintenance activities specifically address energy conservation.

A lighting system maintenance program changes the mode of operation of the facility from one of reaction to one of scheduled maintenance. Scheduling maintenance for lighting systems offers several advantages for both the facility manager and the building occupant. It provides better, more uniform lighting. It permits more efficient use of maintenance personnel and equipment. It helps ensure that the maximum benefit is realized for every lighting energy dollar spent. It reduces overall lighting energy use.

In the past, a number of programs have been implemented to reduce energy used by lighting systems. Tubes were removed. Ballasts were disconnected. Lighting levels were reduced. The objective of these programs was to save energy by reducing the level of lighting provided. While these programs did reduce lighting system energy use, they did little to address the maintenance of the system or the quality of the light produced. Planned lighting system maintenance will do both, while maintaining the appropriate level of lighting for the types of activities that are being performed in the space.

If the goal of the lighting system maintenance program is to provide the recommended level of lighting at a minimum energy cost, there are three things that the program must be designed to accomplish: (1) make use of the most efficient suitable light source; (2) provide the light only where and when it is needed; and (3) take steps to ensure that the light sources are operating as efficiently as possible. Implementing a maintenance program for lighting systems requires an understanding of the fundamentals of lighting and how they relate to lighting system operation.

Lighting Terminology

Lighting, like many other specialized fields, has developed its own terminology. Presented below are brief descriptions of some of the more important lighting terms, presented in a simplified form rather than their more formalized definitions. Most lighting terminology is used to describe two primary features of the lighting system, quantity of light and quality of light.

The lumen is the basic unit of measure for the quantity of light produced by a light source. Since the light output from lamps depreciates as the lamps age, most lamps are rated for both their initial and mean lumens produced. Initial lumens produced is the light output value after the lamp has stabilized in operation, typically after it has operated for 100 hours. Mean lumens is the average light output of the lamp over the rated life of the lamp. The two values can vary significantly. For example, one brand of F32T8 fluorescent lamps has an initial lumen rating of 2,900, and a mean lumen rating of 2,600. Both lumen ratings must be taken into consideration when evaluating lighting requirements for an area.

Since the lumen output of a lamp is a measure of the total light produced by the lamp, it can serve as a measure of the lamp's efficiency or efficacy. Efficacy, calculated in lumens per watt, can be used to evaluate the various efficiencies of the same type of light source, or the relative efficiencies of different light sources. Efficacies range from a low of approximately six lumens per watt for low-wattage incandescent lamps to a high in the range of 60 lumens per watt for low-pressure sodium lamps. Figure 10.1 lists the range of efficacies for common lamp types.

The total lumen output of the lamp does not tell the whole story of quantity of light. While the lumen output of a lamp rates the quantity of light produced by the lamp, it does not give an indication of how much of that light reaches the work surface or the object being illuminated. Illuminance, measured in foot-candles, is the most common measure of the quantity of light that reaches the work surface or the task being performed in the lighted area. One foot-candle of illuminance equals one lumen per square foot. Lighting tables use illuminance values to identify the recommended lighting levels for particular activities.

The most commonly used measure of the quality of the light produced is color temperature. Color temperature is derived from the principle that when an object is heated to a high temperature, it begins to glow. As the temperature of the object increases, the color of its glow changes. If the object were a perfect radiator, there would be a direct relationship between the color of the light radiated and the temperature of the object as measures in degrees Kelvin. Although the color temperature is measured as a single value, the light emitted by the object is actually emitted over a fairly wide spectrum with its center found at that temperature.

That same principle is used to rate the relative color of a light source. The color temperature of the light source in degrees Kelvin gives a method for comparing the distribution of light produced by different types of light sources. Lower color temperatures, typically those less than 3,000 K, have more reddish tones and are considered

Figure 10.1
Efficacies for Common Lamps

Incandescent
> Available in sizes ranging from 5 to 1,500 watts
> Efficacies range between 6 and 22 lumens per watt, generally increasing with wattage
> Rated life of 750 to 1,000 hours

Mercury vapor
> Available in sizes ranging from 40 to 1,000 watts
> Efficacies range between 24 and 60 lumens per watt, increasing with wattage
> Rated life of 12,000 to 24,000 hours

Self-ballasted mercury
> Available in sizes ranging from 160 to 1,200 watts
> Efficacies between 14 and 45 lumens per watt, increasing with wattage
> Rated life of 12,000 to 24,000 hours

Fluorescent
> Available in sizes ranging between 9 and 200 watts
> Efficacies between 55 and 100 lumens per watt, varying with wattage, type of lamp, and type of ballast
> Rated life of 7,500 to 24,000 hours

Self-ballasted fluorescent
> Available in sizes ranging between 5 and 44 watts
> Efficacies between 25 and 60 lumens per watt, varying withw wattage and type of ballast
> Rated life of 7,500 to 12,000 hours

Metal halide
> Available in sizes ranging between 175 and 1,500 watts
> Efficacies between 80 and 125 lumens per watt, varying with wattage
> Rated life of 5,000 to 20,000 hours

High-pressure sodium
> Available in sizes ranging between 35 and 1,000 watts
> Efficacies between 40 and 140 lumens per watt, varying with wattage
> Rated life of 24,000 hours

Color-corrected high-pressure sodium
> Available in sizes ranging between 35 and 100 watts
> Efficacies between 40 and 47 lumens per watt
> Rated life of 24,000 hours

Low-pressure sodium
> Available in sizes ranging between 18 and 180 watts
> Efficacies between 100 and 200 lumens per watt, generally varying with wattage
> Rated life of 14,000 to 18,000 hours

to be warmer. Color temperatures above 4,000 K are bluer and are considered to be cooler. Figure 10.2 lists the color temperature of various light sources.

It should be noted that the color temperature rating system is accurate only for incandescent light sources. Incandescent lamps produce their light across a fairly wide spectrum while electric discharge lamps, such as a metal halide lamp, tend to produce light in narrow bands. The visual effects of incandescent and electric discharge lamps differ widely even though they may have the same color temperature rating. In spite of these limitations, the color temperature rating system is widely used.

Another quality of light measure that is used with nonincandescent lamps is the color renderings index, or CRI. Color rendering indicates how well a light source renders or shows colors. With nonincandescent lamps emitting their energy in narrow bands rather than a continuous spectrum, it often is difficult or impossible to discriminate subtle differences in shades of color of objects. To rate how well the light from a particular light source shows variations in color shades of an object, manufacturers rate the lamp's CRI on a scale of zero to one hundred, with one hundred being daylight. For most applications, a CRI of 60 or higher is considered to be good. Applications requiring a high degree of color discrimination, such as with retail displays, typically use a light source with a high CRI.

Efficiency of Light Sources

One of the most important considerations when establishing a maintenance program for lighting is the type of light source that is used in an application. The choice of light source will have a direct impact on the energy costs of operating the lighting system, and the cost of maintaining the system. The guideline to follow is that the most efficient suitable light source should be used in an application. *Rule of Thumb:* While the efficiency of a particular light source varies by wattage and manufacturer, in comparison to incandescent lamps, mercury lamps use one-third the energy, fluorescent use one-fourth, metal halide use one-fifth, and high-pressure sodium use one-sixth.

Although achieving energy efficiency is one of the goals of the lighting maintenance program, it is not always possible to select the most efficient source as not all light sources are suitable for all applications. Different light sources have different operating characteristics that must be considered, including efficiency, brightness, physical size, wattage, color temperature, color rendering ability, and the time required to restart after a power interruption. Before selecting a particular light source, consider that source's characteristics and how they will impact operations.

Incandescent Lamps

Incandescent lamps are the least cost-effective lighting source widely used in facilities today. With efficiencies ranging from 6 to 22 lumens per watt, they are the lowest efficiency source, typically using four times more energy than their fluorescent counterparts, and five to six times more energy than metal halide lamps. Equally important,

Figure 10.2
Color Characteristics for Common Lamp Types

Lamp Type	Color Temperature (°K)	CRI
Incandescent	2,700-5,000	100
Mercury vapor, standard	5,900	15
Mercury vapor, delux	4,000	50
Mercury vapor, white delux	3,600	60
Fluorescent, warm white	3,000	52
Fluorescent, delux warm white	3,000	77
Fluorescent, cool white	4,100	60-62
Fluorescent, delux cool white	4,200	89
Fluorescent, compact	2,700-5,000	80-89
Metal halide	3,200-4,200	60-80
High pressure sodium	1,900-2,000	9-22
High pressure sodium, color improved	2,300	65
High pressure sodium, color corrected	2,600-2,700	70-80

the lamps have a short life expectancy, typically rated between 750 and 1,000 hours—far shorter than any other type of bulb. The low operating efficiency and the short life expectancy result in both high energy costs and high maintenance costs.

In spite of the limitations of incandescent lamps, they remain widely used in facilities today. They are the least expensive lamp and fixture to purchase. They do not require a ballast in order to operate. They restart instantly following a power interruption. Their operation is not seriously impacted by temperature. They are available in a wide rage of bulb styles and wattages.

Even with all of these advantages, there are very few areas where incandescent lamps should be used today. They are simply too expensive to operate and maintain. Alternative light sources are available for nearly all applications where incandescent lamps are currently being used. The resulting energy and maintenance cost savings typically will pay for their replacement in one to two years.

Mercury Vapor Lamps

Mercury vapor lamps are high-intensity discharge lamps that operate by passing a current through a gas containing metal vapors. Electrons from the current collide with atoms of the metal vapor, energizing the electrons. When the electrons release their acquired energy, they emit light. The wavelength of the light emitted depends on the metal vapor used. A current limiting ballast is required for the lamps to operate. Most ballasts are externally mounted, although a few models are available with built-in ballasts.

Most mercury vapor lamps are used in outdoor applications. Developed to overcome problems with starting fluorescent lamps at low temperatures, the lamps are more efficient than incandescent lamps but less efficient than fluorescent. Typical lamp efficiencies range between 24 and 60 lumens per watt for units with external ballasts. Efficiencies for self-ballasted lamps are 10 to 20 percent lower. It is this relatively low efficiency that makes mercury vapor lamps candidates for replacement with newer, higher-efficiency light sources as part of a maintenance program.

Mercury vapor lamps are available in sizes ranging from 40 to 1,000 watts, with higher-wattage lamps offering higher efficiency. By varying the composition of the phosphors used in making the lamps, manufacturers have developed a range of different color temperatures. The three most common of these are: (1) 5,900 degree Kelvin standard white; (2) 4,000 degree Kelvin delux; and (3) 3,600 degree Kelvin white delux coatings. Color rendering falls in the range of 20 to 60, providing fair to good rendering. Lamps have a rated life of between 16,000 and 24,000 hours.

One drawback of mercury vapor lamps is that they require between five and seven minutes to restart and achieve full light output following a power interruption. As a result, their use is restricted to applications where this restarting delay is acceptable.

One of the most significant drawbacks of mercury vapor lamps is their poor lumen maintenance. As the lamps age, their light output steadily declines. By the time the lamps reach their rated life, their light output will have decreased below their rated output by as much as 25 percent. Since mercury vapor lamps rarely fail even when they reach their rated life, many continue to operate with declining light output. It is not uncommon to find lamps in operation that produce less than 50 percent of their rated light output.

Most manufacturers recommend that a group relamping program be established for mercury vapor lamps in order to maintain a minimum efficiency level for the lighting system. Relamping intervals can be based on measured light output or hours of operation.

Fluorescent Lamps

The fluorescent lamp is the lighting workhorse of facilities. Originally developed during the late 1930s and early 1940s as an energy efficient alternative to the incandescent lamp, fluorescents have continued to evolve. Today, fluorescent lamps are available in a wide variety of lamp shapes and sizes and are used in settings ranging from hallway and office lighting to retail applications. The popularity of fluorescent lamps is such that it would be difficult to find a facility that does not make extensive use of at least one type of fluorescent lamp.

All fluorescent lamps require the use of a ballast. Generally the ballast is mounted externally, although some compact fluorescent lamps have built-in ballasts. The standard magnetic ballast is rapidly being replaced with electronic ballasts that promise increased energy efficiency and improved operation of the lamp through reduced flicker. Special low-temperature ballasts are available for outdoor applications to help overcome problems with lamp starting at low temperatures.

Standard fluorescent lamps are available in wattages ranging from 9 to 200. Efficiency varies with both wattage and manufacturer, falling in the range of 55 to 100 lumens per watt. Most lamps have a rated life of between 7,500 and 24,000 hours, depending on the type and wattage of the lamp.

Manufacturers offer a range of different lamp colors by varying the mixture of phosphors used to coat the inside glass envelope of the lamps. The two most popular colors are cool white and warm white. Standard cool white lamps operate at 4,100 degrees Kelvin while warm white lamps operate at 3,000 degrees Kelvin. Both lamps have a CRI rating in the low 60s. Lamps that make use of special phosphor coatings, such as the triphosphor lamp, have CRI ratings as high as 85.

Compact fluorescent lamps are a relatively new type of fluorescent lamp. Designed as a direct replacement for the incandescent lamp, most are self-ballasted and can be installed directly in the incandescent fixture without modification. Efficiency varies with lamp design, but they typically produce three to four times the lumen output for the same wattage incandescent lamp. Most have a color temperature between 2,700 and 4,100 degrees Kelvin and a CRI in the low 80s. Although their rated life of 12,000 hours is less than for standard fluorescent lamps, they last 12 to 15 times as long as the incandescent lamp they are designed to replace.

Metal Halide

Metal halide lamps were developed from the basic mercury vapor lamp design in an attempt to improve on their efficiency. The result is a lamp that is nearly twice as efficient at 80 to 125 lumens per watt, produces a whiter light, and has better color rendering capabilities. Like the mercury vapor lamps, metal halide lamps operate by producing an electric arc and require the use of a current limiting ballast.

The color properties of metal halide lamps make them well suited for applications where distinguishing colors is important, such as in retail applications, grocery stores, stadiums, and warehouses. Most metal halide lamps operate with a color temperature in the range of 3,200 to 4,200 degrees Kelvin, with a CRI between 60 and 80. Improved color lamps are available for applications requiring even better color rendering. Some of these lamps operate with a rated CRI in excess of 80.

Lamp wattages range from 175 to 1,500 watts. Lamp life ranges from 5,000 to 20,000 hours, depending on the wattage. Metal halide lamps, however, have a fairly high rate of lamp lumen depreciation; higher than most other lamp types. Typical lamp light output falls to approximately 70 percent of the initial light output when lamps reach their rated life. The color of the light produced also changes with lamp age, making implementation of a group relamping program a necessity.

Metal halide lamps require a fairly long restrike time following a power outage. Before the lamp can be restarted, it must cool, resulting in a delay of ten to fifteen minutes.

High-Pressure Sodium

High-pressure sodium lamps are high intensity discharge (HID) units that contain a mixture of vapors including mercury and sodium. Three types of lamps are produced—standard, color-improved, and color-corrected. Standard lamps produce a golden-white light that enhances yellows and oranges while graying blues and subduing reds and greens. These lamps operate with color temperatures in the range of 1,900 to 2,000 degrees Kelvin, with a CRI in the low 20s. Lamps are available in wattages ranging from 35 to 1,000, with efficiencies of 40 to 140 lumens per watts depending on lamp wattage. Standard high-pressure sodium lamps are used primarily in applications such as street lighting, hi-bay industrial lighting, greenhouses, and outdoor floodlighting where color perception is not important.

Color-improved high-pressure sodium lamps operate at approximately 2,300 degrees Kelvin with a CRI in the mid-60s. This improvement in color properties does cause a slight decrease in efficiency. Efficiencies for color-improved lamps range from 60 to 90 lumens per watt, depending on lamp wattage. Lamps are available in sizes ranging from 70 to 400 watts. Although color-improved lamps are an improvement over the standard lamp, color consistency is a problem resulting in limited use in applications beyond those for the standard lamp.

Color-corrected high-pressure sodium lamps offer the best color characteristics, approaching those of incandescent lamps. Color-corrected lamps operate at between 2,600 and 2,700 degrees Kelvin with a CRI rating in the mid-80s, making them well suited for applications requiring excellent color rendition, such as retail facilities. The improvement in color characteristics comes with an efficiency penalty. Efficiencies range from 40 to 47 lumens per watt for bulb sizes of 35 to 100 watts. Even at these lower efficiencies, color-corrected high-pressure sodium lamps offer three times the efficiency of incandescent lamps.

Of all the HID light sources, high-pressure sodium lamps offer the best lumen maintenance. The lamps maintain approximately 90 percent of their initial lumens throughout their estimated 24,000–hour life. Like other HID light sources, high-pressure sodium lamps must cool after a power interruption before they can be restarted. Most restart times range between five and ten minutes

Low-Pressure Sodium

Low-pressure sodium lamps are the most efficient, commonly used light source available today. Efficiencies range as high as 200 lumens per watt. In spite of their high efficiency, their use is limited primarily to outdoor lighting applications, such as parking lots, highways, and security lighting because of their poor color characteristics. About the only indoor application is for a limited number of warehouses where color perception is not important.

The light produced by low-pressure sodium lamps is concentrated in two narrow bands of the spectrum, resulting in all color other than yellows being misrepresented as various shades of gray or black. The expected life of the lamps ranges between 14,000 and 18,000 hours. Throughout its rated life, the light output of the lamp remains nearly constant, decreasing by less than 10 percent by the time the lamp reaches its rated life expectancy. However, the current draw of the lamp increases as the lamp ages, resulting in an ideal candidate for a group relamping program in order to reduce maintenance and energy costs.

Low-pressure sodium lamps will restart immediately following a power interruption. Most restart without cooling, although they will take between five and fifteen minutes to achieve full light output.

HOW TO DETERMINE LIGHTING REQUIREMENTS ————————

There are three primary factors that determine how much light is required for performing a particular task: (1) the average age of the people who are performing the task; (2) the importance of speed and accuracy; and (3) the reflectance of the task background. Since these factors vary with individual applications, most tables of recommended lighting levels list a range of values for specific tasks rather than a single value. Understanding how these factors impact the need for lighting will help in developing a maintenance program that meets the needs of the facility occupants while keeping energy and maintenance costs as low as possible.

Worker Age

As people age, their eyes change both in their ability to focus and in their efficiency. As a result, the average person by age 60 requires nearly 25 percent more light than a 20-year-old needs in order to achieve the same visual acuity. Lighting maintenance programs must take the average age of the occupants into consideration if the lighting system is to provide sufficient illumination. Since the age range of occupants in a particular area may be very wide, it will be necessary to compromise on the actual lighting level provided. Some occupants will have more light than they require, while others will have slightly less. The key is to find and maintain a balance that meets everyone's needs.

Speed and Accuracy

Higher lighting levels allow workers to distinguish objects and tasks with a decreased chance for error. When lighting levels are too low, visual tasks, particularly those requiring high contrast or those that must be performed for an extended period of time, become even more difficult. Increasing the lighting level will improve visibility for workers, but there is a limit. Lighting levels that are too high are just as detrimental to performance as are levels that are too low. The best performance will occur within the recommended range of illumination.

Contrast

The level of reflectance of the task background provides contrast for the worker. The greater the contrast, the easier it is to see the task being performed. If the contrast is poor and cannot be improved due to the nature of the task being performed, then lighting levels will have to be increased to improve performance.

Recommended Lighting Levels

Figure 10.3 lists the Illuminating Engineering Society's recommended ranges for illumination in foot-candles for various tasks. Three values are given for each application: low, medium, and high. For most activities, the target value for the lighting system is the middle of the three values. The higher value is the upper range for that task based on the particular conditions found in that application.

In practice, if lighting systems were properly maintained, most areas would be overlit. Designers build in higher-than-required lighting levels to compensate for losses as the system ages as well as for changes in the layout of the space. Lamp lumen depreciation and dirt accumulation alone can reduce lighting levels by as much as 30 percent in as little as two years. Proper maintenance will counter the impact of both of these factors, and will achieve energy savings by allowing the total number of fixtures in a given space to be reduced, or by allowing lower-wattage lamps to be installed.

Figure 10.3
Recommended Illumination Levels

Activity	Illuminance (fc)	Typical Applications
Public spaces with dark surroundings	2 - 3 - 5	Night lighting of hallways
Simple orientation for short temporary visits	5 - 7.5 - 10	Restaurant dining areas, inactive storage areas
Working spaces where visual tasks are only occasionally performed	10 - 15 - 20	Auditoriums, elevators, lobbies, hallways, lounges
Performance of visual tasks of high contrast or large size	20 - 30 - 50	Conference rooms, library book stacks, cashier areas, hotel bedrooms, warehouse areas, simple assembly
Performance of visual tasks of medium contrast or small size	50 - 75 - 100	Mail sorting, reading poor copy, drafting, classrooms
Performance of visual tasks of low contrast or very small size	100 - 150 - 200	Fitting rooms, proofreading, difficult assembly
Performance of visual tasks of low contrast and very small size over a prolonged period	200 - 300 - 500	Inspection, very difficult assembly
Performance of very prolonged and exacting visual tasks	500 - 750 - 1,000	Exacting assembly, very fine bench work
Performance of very special visual tasks of extremely low contrast and small size	1,000 - 1,500 - 2,000	Surgical procedures

Note: Courtesy of the Illuminating Engineering Society of North America, 120 Wall Street, 17th Floor, New York, NY 10005.

Before fixtures can be removed or downgraded, it will be necessary to survey the current lighting systems in the facility. The survey should collect information on the lighting levels provided by the system, the number and type of light sources installed in the space, and the type of controls installed for the system. Use Figure 10.4 to record lighting system information important to the operation and maintenance of the lighting system.

Lighting surveys, while not difficult, are very time consuming. It takes time to identify, collect, and record the necessary data. Special precautions must be taken in order to ensure that the readings taken are accurate. All effects of daylight must be eliminated. The lights will have to be at the proper operating temperature in order to provide accurate information. Even fluorescent lamps take several minutes to stabilize at their proper operating temperature. HID systems will take as long as fifteen minutes. Taking readings before the system has properly stabilized will result in incorrect values.

Another important consideration before conducting the lighting survey is the condition of the system. If there has been no planned maintenance program for lighting, levels will be lower than planned as a result of lamp lumen depreciation and an accumulation of dirt on the lamps and fixtures. Discolored lenses may also block a significant portion of the light given off by the fixtures. Therefore, before meaningful light readings can be taken, it will be necessary to replace all discolored lenses, clean the lenses and fixtures, and relamp the fixtures with new bulbs. By taking these actions, you can be assured that the light reading that will be taken will reflect a system that is being properly maintained.

HOW TO DEVELOP A LIGHTING MAINTENANCE PROGRAM

The primary purpose of a lighting maintenance program is to ensure that the lighting system consistently provides the proper quantity and quality of light required for the tasks being performed at the lowest possible energy cost. Selecting the most appropriate light source is the first step in implementing the program. Unfortunately, few facilities routinely examine the question of changing the light source in order to improve system performance, energy efficiency, and maintenance requirements. Most are simply content with replacing burned out lamps.

Since the selection of the light source has such a large impact on performance and operating costs, you must regularly review the particular systems installed and consider if there are more appropriate, efficient, or lower-maintenance alternatives.

To ensure the system operates at its peak efficiency and that energy use of the system is minimized, four additional steps are required: (1) limit the operation of the lights to only when they are needed; (2) implement a program for scheduled replacement of the lamps; (3) regularly clean the lamps and fixtures; and (4) inspect lighting system components on a regular basis.

Step 1: Use Automated Lighting Controls

The purpose of automated lighting controls is to ensure that the lights are used only when they are actually needed. By limiting the operation of the lighting system, energy

Figure 10.4
Ligthing Survey

1. Building: _____

2. Room/area: _____

3. Area (sq ft): _____   4. Ceiling height: _____

5. Wall color: ☐ light   ☐ medium   ☐ dark

6. Visual task being performed: _____

7. Average lighting level (fc): _____

Lighting Fixture Type 1

1. Number of fixtures: _____   2. Number of lamps/fixture: _____

3. Type of lamp: _____

4. Watts/lamp: _____   5. Type of diffuser: _____

6. Type of control: _____   6. Overall condition: _____

Lighting Fixture Type 2

1. Number of fixtures: _____   2. Number of lamps/fixture: _____

3. Type of lamp: _____

4. Watts/lamp: _____   5. Type of diffuser: _____

6. Type of control: _____   6. Overall condition: _____

Lighting Fixture Type 3

1. Number of fixtures: _____   2. Number of lamps/fixture: _____

3. Type of lamp: _____

4. Watts/lamp: _____   5. Type of diffuser: _____

6. Type of control: _____   6. Overall condition: _____

use and maintenance requirements will be reduced. Lighting controls are the most cost-effective element in any lighting maintenance program. Their benefits are concrete, easily seen, and readily quantified. Automatic lighting controls typically pay for themselves in energy savings in one year or less.

How effective are lighting controls in reducing operating costs? While the actual savings achieved will vary with the installation, typical results show that an automated lighting control system will reduce lighting energy use by one-third. Maintenance costs as a result of reduced operating time will also be reduced by approximately 25 percent.

The need for lighting controls can be easily identified simply by walking through the facility while it is occupied and again while it is unoccupied. During normal business hours, note how many areas have their lighting systems turned on when there is no one present. Offices, conference rooms, storage, and other spaces typically remain unoccupied for hours at a time, yet the lighting system is left on. As a result of time-of-day rate schedules, the system is using energy when it is the most expensive, and adding to the facility's peak demand charges.

During the air conditioning season, unnecessary daytime lighting system operations also impact the load on the air conditioning system. Every watt of energy used by the lighting system ends up as heat introduced into the conditioned space—heat that must be removed by the air conditioning system. Reducing the use of lighting systems to when they are actually needed reduces the load on the air conditioning system, saving additional energy.

The unoccupied hours walk-through is equally important. While some lighting is required to support after-hours operations and to enhance security within and around the facility, in many cases lighting is left operating after hours either because occupants failed to turn it off or because there are no means to turn it off. For example, it is common practice in many office and institutional facilities to leave the hallway and corridor lighting operating while the facility is unoccupied. In some cases, the lighting is wired directly to a circuit breaker with no manual on–off control. With these facilities unoccupied for 50 percent or more of the time on a weekly basis, the energy and maintenance savings that can be achieved by turning the lights off will typically pay for the cost of installing the necessary controls in a matter of months.

There are two classes of automated controls for lighting systems: on–off controls and variable lighting output controls. Both can be used effectively in facilities to reduce lighting energy use and maintenance costs. Of the two, on–off controls are simpler to implement, less expensive to install, and provide the greatest rate of return.

Automated on–off controls are best suited for applications where lighting is not needed for even rather short periods of time, such as private offices, conference rooms, classrooms, and hallways. The three most common methods of implementing on–off control are through the use of time clocks, occupancy sensors, and remotely programmed controls. All three types reduce the maintenance requirements of the lighting system by reducing the hours of operation, and all three save energy. However, not all three are suitable for all applications. Each has its advantages and limitations that must be considered before applying them to a particular application.

Time clocks, including electromechanical and electronically based units, are the simplest, least expensive, and most commonly used method for controlling lighting systems. Best suited for applications where lighting is needed on a fixed schedule, such as exterior lighting, interior and exterior security lighting, and large interior areas, time clocks provide a simple means of ensuring that the lighting system is turned off when the lights are no longer needed. Time clocks may be used with any of the light sources commonly found in facilities. Their most serious limitation is the need to reset their timers when operating schedules change, or when a power outage interrupts their operation.

Occupancy sensors are an effective means of reducing lighting energy use in spaces where there is no set pattern of use, yet there is a need to limit the operation of the lighting system to times when someone is present in the area served by the lights. Classrooms, private offices, and conference rooms are examples of areas that are frequently unoccupied for hours at a time even though the facility itself remains occupied. Since the time that they are unoccupied varies from day to day, it is impossible to make use of fixed time clocks to limit the operation of the lights.

Occupancy sensors detect motion in the area served by the lighting system through the use of infrared or ultrasonic sensors. When the sensor fails to detect someone in the space, it turns the lights off. To prevent the turning off of lights when the space is occupied but there is little motion to be picked up by the sensors, the units have an adjustable time delay ranging from one to sixty minutes. Units are available as direct replacements for standard wall switches for small areas, or as ceiling and wall mounted units for larger areas. They are suitable for use with only incandescent and fluorescent lamps. Occupancy sensors cannot be used with HID light sources due to the lengthy warm-up time requirements.

Although occupancy sensors can reduce lighting energy use by 25 to 35 percent over manual controls, some facility managers have been reluctant to install them because of their impact on lamp life, particularly for fluorescent lamps. With manual wall switches, the lamps are started only a few times each day. With occupancy sensor controls, the potential exists for starting lamps much more frequently each day. Since lamp life is related to the number of times the lamps are started, lamp life will be decreased in a system that uses occupancy sensor controls.

Although occupancy sensors will decrease the life of the lamps, the cost of the lost lamp life is small in comparison to the energy savings produced. For example, a 40-watt fluorescent lamp uses 960 kWh of electricity over its 24,000-hour rated life. At a rate of 10 cents per kWh, the cost of the electricity used by the lamp over its life is $96. Assuming that the more frequent starting of the lamp reduces its life by 50 percent, the lamp controlled by an occupancy sensor will use approximately $48 worth of electricity. With occupancy sensors typically reducing lighting energy use by a minimum of 25 percent, the savings produced by turning the light off more frequently is at least $12; sixteen times the cost of the lost lamp life.

Remotely programmed controls constitute a third method of providing automated on–off lighting system control, such as that found in a central energy management system. Remotely programmed controls function similar to time clock controls, only with the ability to vary the time settings remotely. Further, the system operator can readily override the time controls and bring the lights on when there is a needed deviation from the programmed schedule, or can turn them off when there is a need to reduce electrical demand.

The alternative to the on–off control is the variable lighting control. Variable lighting controls save energy by making use of daylight to meet part of the lighting requirements. A series of photocells measure the total light available in the space. As the quality of daylight available increases, the system automatically cuts back on the amount of lighting produced by the fixtures in order to maintain a constant level of illumination.

Variable lighting control systems are currently used with only incandescent and fluorescent lamps. For incandescent lamps, no special equipment other than the controls is required. For fluorescent systems, special dimming ballasts must be installed that are then controlled by the variable lighting control system. While other light sources can be used, the cost of their special ballasts generally makes their use too expensive.

Although on–off controls are simpler and less expensive to install than variable lighting control systems, there are applications where on–off controls are not suitable. For example, in facilities with a large quantity of glass, perimeter areas are overlit most of the time. While on–off lighting control systems could be used to reduce the lighting load in these areas, they only offer two, or at most, three, levels of lighting, resulting in discrete jumps in lighting levels as the lamps are turned on and off. Variable lighting control systems offer infinite levels based on the available sunlight.

Step 2: Schedule Lamp Replacement

Most facilities replace lamps on a spot basis after a lamp has failed. Maintenance personnel are informed of the location of the failed lamp, identify the type of lamp used, get a ladder and the required lamps from the storeroom, carry them to the location, replace the lamp, return the ladder, and discard the failed lamp. In a few facilities, the fixture may be cleaned during the spot relamping process. Studies of facilities that spot relamp show that the time involved, including travel and set-up time, averages nearly twenty minutes per lamp. In contrast, if all of the lamps in an area were replaced at the same time, the time involved per lamp would be reduced to between two and five minutes, including the time required to clean and inspect the fixture.

Another problem with the practice of replacing lamps only on a spot basis when they burn out is that lighting levels suffer. The lumens generated by all lamps degenerate with time. Replacing lamps on a spot basis when they fail results in an increase in the average age of the lamps. As their age increases, their light output decreases, resulting in significantly lower lighting levels although energy use by the lighting system does not decrease. The result is a lighting system that does not operate very efficiently. For example, a fluorescent lighting system that is spot relamped and does not undergo a regular schedule for cleaning may produce only 50 to 80 percent of the light output that it did when it was new.

A third problem caused by spot lamp replacement is uneven lighting levels. When individual lamps are replaced, the light output of that fixture is increased to nearly what it was when the system was new, particularly if the fixture is cleaned when the lamps are replaced. In contrast, other fixtures in the area that were not relamped will be producing only 50 to 80 percent of the light output of the newly relamped fixture. The result will be noticeably uneven lighting levels.

Finally, spot replacement programs can increase the wear and tear on components such as lamp ballasts. When a lamp fails, the ballast typically keeps on trying to start. This process increases the operating temperature of the ballast, and can lead to early ballast failure. Under spot replacement programs, it may be days or weeks before the lamp is actually replaced.

The alternative to spot lamp replacement is a program that schedules the replacement of all lamps in an area at the same time. Group relamping is not new. Lighting organizations have been promoting the benefits of group relamping for years, including more energy efficient lighting, lower labor costs, better maintained lighting levels, improved appearance, fewer interruptions for the building occupants, and less wear on lamp ballasts.

In spite of these benefits, few organizations practice group relamping. One of the leading factors that has prevented organizations from implementing group relamping programs is that maintenance personnel simply find it difficult to throw away lamps that are operating. What they fail to realize is that while those lamps are operating, they are no longer operating efficiently. Lamp lumen depreciation is gradual and nearly impossible to notice. But it is real and can significantly impact the operation and efficiency of the lighting system.

Another factor contributing to the reluctance to implement group relamping programs is their perceived cost. When lamps are replaced on a spot basis, their cost is spread out over a long period of time. Spot relamping labor costs are generally soft costs since they are performed by in-house personnel and are hidden among the other activities performed by the personnel. In contrast, group relamping costs are hard costs that can be readily identified. In group relamping programs, the cost of the lamps is concentrated in large sums of money that must be spent every few years. Further, since some of the remaining life of the lamps is being discarded, the quantity of lamps that will be required over a given period of time will increase, further increasing lamp costs. The labor costs are readily identified, either through the use of a contract or a dedicated maintenance crew.

What is not realized is that the cost of the lamp is a very small part of the cost of relamping. For example, if it takes an organization approximately 20 minutes to replace a lamp on a spot replacement basis, and the labor costs are $15 per hour, the labor charge for that one lamp replacement is $5. Assuming a lamp cost of $1.50 per bulb, the total cost of spot relamping is $6.50 per bulb, with no fixture cleaning.

Studies of facilities that use group relamping have shown that the average time required to relamp and clean a fixture averages between two and three minutes per lamp. Assuming three minutes per lamp replacement time, and the same labor rates and lamp costs, the total cost to replace lamps under a group relamping program is $2.25 per lamp, a savings of $4.25 per lamp over spot replacement costs. And remember, the group relamping labor costs include the cost of cleaning the fixture, something that is not included in spot replacement costs.

To make the most effective use of group relamping programs, you must set the relamping interval properly. Every lamp manufacturer publishes lamp lumen depreciation and mortality curves for their products. Figure 10.5 shows a survival rate for a common fluorescent lamp. From the figure it can be seen that at 100 percent of the rated life of the lamps, approximately 50 percent of them will have failed. At 70 percent of their rated life, less than 15 percent of the lamps will have failed. Actual rate of failure will vary with the number of times that the lamp is started during this period. After the lamps have reached 70 percent of their rated life, the number of lamp failures will increase rapidly, as shown in the figure.

Most manufacturers of commonly used fluorescent lamps recommend that the group relamping interval be set to when the lamps have reached approximately 70 percent of their rated life. At this point, the lamps will still have a good lumen per watt rating as their lamp lumen depreciation rate is slow. Again, the actual interval will depend on the number of hours the lamps burn per start.

Setting the relamping interval is a balance between lamp lumen depreciation, lamp failure rate, and the expected life of the lamp. The goal of the program is to replace the lamps before their light output has decreased significantly and before there are too many lamp failures. Selecting the right interval will maintain a more uniform lighting level, decrease lamp replacement costs, and will enable you to set lower initial lighting levels, knowing that the lamps will be replaced before their lumen output decreases significantly.

While a target relamping interval of 70 percent of rated life is a good one, it is difficult to realize. Unless all of the lamps burn exactly the same number of hours per start, it will be impossible to determine when they have reached 70 percent of their rated life. While it is possible to identify when lamps have reached their 70 percent life expectancy level by keeping records of the number of lamps that burn out, and group relamping when 15 percent have failed, this is generally not practical. But there are alternatives.

One guideline that can be followed to help determine the best time for relamping is based on the number of average hours that lamps burn per start. If lamps are permitted to burn 24 hours per day, relamp every year. If lamps burn on the average 16 hours per start, relamp every two years. If lamps burn only 8 hours or less per start, relamp every three years.

Another method that can be used to identify the ideal relamping interval is to retain approximately 15 percent of the lamps removed during a group relamping effort. As lamps fail after the group relamping, these retained lamps are used as replacements. When the number of retained lamps is nearly depleted, the area is ready to be group relamped once again. If this method is used, care should be taken to ensure that only the lamps that show the least amount of wear are retained, and that the lamps that are retained are used only in the designated area.

Step 3: Clean Lamps and Fixtures

Cleaning lamps and fixtures is an important part of any lighting maintenance program. Cleaning helps by maximizing the reflectance of fixture surfaces and by minimizing

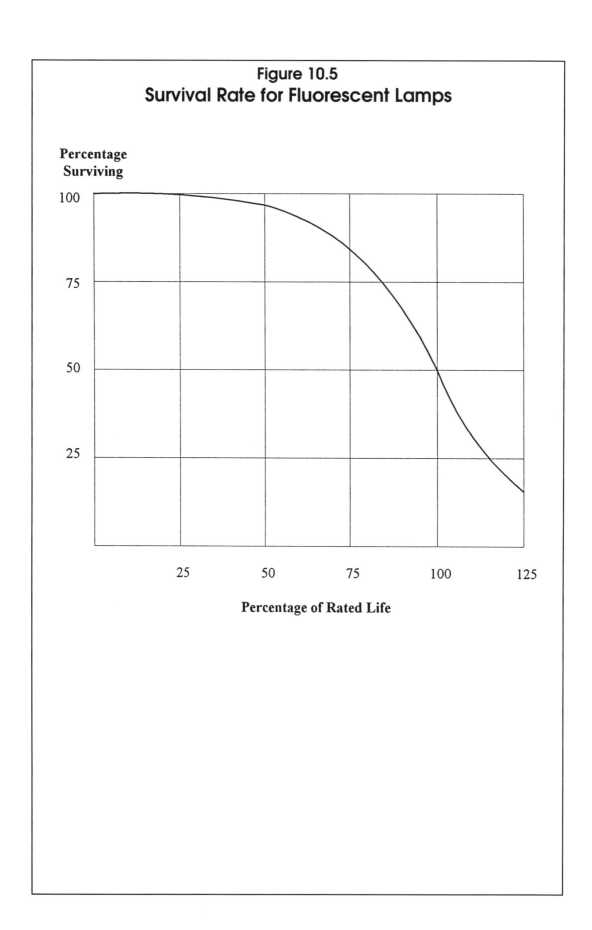

Figure 10.5
Survival Rate for Fluorescent Lamps

the blocking of light due to the presence of dirt on the lamp, the lens, the fixture shielding, and the fixture surfaces. Few facilities, however, have a comprehensive lamp and fixture cleaning program in place. In most facilities, if the fixtures are cleaned, they are cleaned when the lamps are spot replaced.

All lamps and lighting fixtures accumulate dirt over time. The rate of accumulation depends on the type and style of the fixture and the environment in which it operates. Fixtures located in dusty and dirty environments, such as those found in industrial and warehouse facilities, will accumulate dirt much more quickly than those found in typical office environments. But even those found in even relatively clean environments will accumulate dirt. For example, it is not uncommon to find the light losses from dirt accumulation as high as 20 percent from fixtures as little as two years old located in an office environment. In dirtier environments, light loss due to dirt accumulation can be as high as 50 percent.

Although dirt decreases the amount of light available from a fixture, the light continues to use the same amount of energy, resulting in a decrease in lighting efficiency. The net effect is that you are not getting all of the light that you are paying for. The longer the system operates without being cleaned, the more the inefficiency grows. Spot cleaning fixtures when their lamps burn out will not solve the problem due to the long life of most lamps other than incandescent. Spot cleaning also results in uneven lighting levels.

The only way to prevent an accumulation of dirt from gradually reducing the lighting level is to implement a scheduled cleaning program for all fixtures. Scheduled group fixture cleaning is also more efficient than spot cleaning. In nearly all applications, the effort to clean light fixtures can benefit from the use of specialized equipment, equipment that is practical only with group relamping efforts. Lifts, platform ladders, movable scaffolds, and floor carts with tanks for cleaning solutions and rinsing reduce time and labor requirements. If fixtures are cleaned on a spot basis, such as when individual lamps are replaced, the purchase of this equipment cannot be justified.

Scheduling the cleaning of all fixtures in a given area at the same time also minimizes the disruption to operations taking place within the space. Most facilities schedule fixture cleaning activities after normal working hours, thus allowing easier access to the space, as well as minimal disruption to operations. With the use of special cleaning equipment, most areas can be cleaned in a single evening.

To gain the greatest benefit from scheduled fixture cleaning programs, the cleaning interval will have to be based on conditions found in the facility. Dirty environments require shorter intervals between cleanings. While cleaning fixtures once every several years may be sufficient for office environments, dusty and dirty environments will require shorter intervals between cleanings.

A program for lighting fixture cleaning is most cost-effective when it is combined with a group relamping program. The crew and equipment are already in place and can readily perform both tasks at the same time. In clean environments, it will be necessary to perform an additional cleaning at the mid-point between relamping intervals. In dirty environments, it may be necessary to perform two or three cleanings between relamping intervals. Monitoring the decline in lighting levels in the space and correcting for lamp lumen depreciation will help to identify the optimum interval for cleaning.

Step 4: Inspect System Components

The final element of a lighting maintenance program is the inspection and maintenance of the various components in the lighting system. Heat and age are the biggest causes of fixture component deterioration. Reflective surfaces oxidize and discolor, reducing their reflectivity. Lenses discolor and cloud, blocking a significant portion of the light produced by the lamp. Ballasts fail, causing lamps to flicker or to not start at all. Sockets can crack or break during lamp replacement, causing loose connections with the lamp. The result is a slow deterioration of the quantity and quality of light produced by the system and more frequent service calls due to lamp failures.

Regularly scheduled inspections of lighting system components identify problems as they are developing and allow maintenance personnel to correct them before they result in an interruption of service. Equally important, regular inspections identify areas where the quality and quantity of the light being produced by the system is declining so that corrective action can be taken before lighting levels fall too low.

Ideally, the best time to perform the inspection is when the fixtures are being cleaned and the lamps are being replaced as part of a group relamping program. The inspection can be completed by the lighting maintenance crew as the fixtures are opened for cleaning and lamp removal. Minor problems, such as cracked sockets and bad ballasts, can be corrected on the spot, while more extensive problems, such as deteriorated lenses and deteriorating reflecting surfaces, can be noted for future maintenance efforts.

The inspection program should also include a check of the voltage being supplied to the lighting system to verify that the system is operating at its proper voltage. Higher than rated voltages decrease lamp and ballast life while lower than rated voltages decrease light output and make starting some type of lamps more difficult.

IMPLEMENTING THE MAINTENANCE PROGRAM

Implementation of a maintenance program for lighting systems carries a relatively high initial cost in that investments must be made in cleaning equipment, equipment to ease access to lighting fixtures, and a large purchase of lamps to be used in the group relamping effort. In many facilities, at least some of the required equipment will already be available, although it may not be possible to dedicate its use to the lighting maintenance program. Remember that although these costs may seem large, facilities that are currently replacing lamps on a spot basis are already spending more each year for lighting maintenance. The only difference is that those costs are not readily apparent.

From the lighting survey, identify areas where group relamping and cleaning will provide the most benefit in terms of reduced maintenance costs and improved lighting performance. High bay areas, large open office complexes, areas with obstructions—all are very labor intensive areas for spot replacement of lamps and thus offer the greatest return for the investment. Start with these areas, gradually expanding the program to include additional areas. As the area of the facility that is being group relamped increases, the number of maintenance calls for spot replacement of burned out lamps will decrease, further freeing maintenance personnel to participate in the program.

SUMMARY

A maintenance program for building lighting systems offers a very large potential for reducing both the maintenance and energy costs of building lighting systems while improving the overall performance of the systems. Typical energy savings resulting from a comprehensive maintenance program range between 25 and 33 percent. Maintenance labor savings typically range between 50 and 90 percent. Although the programs are labor intensive during start-up, start-up costs are typically recovered through maintenance and energy savings within one year or less.

Chapter 11

Electric Motors

This chapter presents guidelines and procedures on maintaining the types of electric motors commonly found in commercial, institutional, and industrial facilities. The information presented in this chapter is targeted toward the most common type of electric motor found in facilities today, the induction motor. However, nearly all of the maintenance activities and procedures identified may be applied to all types of electric motors regardless of application or size. By following the procedures outlined here, you will be able to establish a maintenance program for your facility's motors that will decrease the number of unscheduled outages resulting from motor failures and decrease the energy requirements of the motors.

The procedures identified in this chapter are intended to be used to establish a scheduled maintenance program for motors supporting typical operations in these facilities. They may have to be modified in terms of the activities performed and the frequency with which they are performed for very large motors and those that support critical operations.

As with most areas of facility maintenance, a terminology specific to motor operation and maintenance has evolved. Throughout this chapter, some of the more commonly used motor-specific terminology will be explained to convey a general understanding of what the terms mean and why they are important. The explanations given in this chapter are informal and are not meant to replace the precise definitions as given by the National Electrical Manufacturers Association (NEMA) or others. If you require a more formal definition or explanation, refer to NEMA Standard MG-1, *Motors and Generators*.

ELECTRIC MOTOR ENERGY MAINTENANCE CONSIDERATIONS

It would be practically impossible to find a facility today that does not make widespread use of electric motors in support of its operations. Ranging from fractional horsepower units, to units rated at 100 horsepower and more, electric motors drive numerous building loads, including pumps, chillers, fans, compressors, elevators, and conveyors. In fact, there are so many electric motors in operation in facilities today that their energy use accounts for slightly more than one-half of the total electricity used annually in the United States American businesses spend an estimated $75 billion in energy costs to operate electric motors each year.

With such a large component of electricity use going to drive electric motors, the potential for energy savings is tremendous. Even a small increase in operating efficiency translates into large savings in electricity costs. For example, a 2 percent improvement in the operating efficiency of all currently installed electric motors would translate to a 1 percent decrease in the total demand for electricity, and a $1.5 billion savings for business.

On a smaller and more understandable level, consider the impact that a 2 percent improvement in operating efficiency would have on the energy used by a 30 horsepower motor driving a fan in an air handling system, or a pump in a chilled or hot water distribution system. Assuming the motor operates for ten hours per day, 250 days per year, at an average electrical rate of 6 cents per kilowatt-hour, a 2 percent improvement in motor efficiency would result in an annual savings of slightly more than $100 per year. Over the life of the motor, the savings would total more than $2,500. The savings would be nearly three and one-half times larger if the motor operated for 24 hours per day, a practice common in central heating and cooling plants and many high-use facilities.

Motor Efficiency Considerations

An electric motor's efficiency is expressed as the ratio of the motor's power output divided by its input power. It is a direct measure of how effectively the motor converts electrical energy into mechanical energy. Since the efficiency of a particular motor varies with the load placed on it, all motor efficiency ratings are for full-load operating conditions.

The difference between the power produced by a motor and its input power are the motor losses. Controlling motor losses through motor design and material modifications and good maintenance practices is the key to improving motor efficiency.

There are three major types of losses that occur in motors: copper losses, iron losses, and mechanical losses. Copper losses are caused by the electrical resistance in the motor windings. They vary with the motor load and are proportional to the square of the motor current. Manufacturers reduce copper losses by increasing the size of the conductors used in winding the motor's coils. Copper losses can be reduced if you ensure that motors operate at the proper voltage, and that the voltage is balanced across the power line phases supplying the motor.

Iron losses are caused by the magnetic resistance of the laminated cores of the stator and rotor. They are independent of the load placed on the motor and can be minimized only through design changes, including the use of high-grade steel and thinner laminations.

Mechanical losses are caused by friction in the rotating components of the motor.

Motor bearings, brushes, and fans are the major sources of mechanical losses. They are independent of load and current, but vary with the temperature of the motor. Manufacturers minimize mechanical losses primarily through efficient cooling fan design. Maintenance personnel can minimize mechanical losses through good lubrication practices and ensuring that the motor is operating at the proper temperature.

Rule of Thumb: (1) As the horsepower rating of the motor increases, so does its full-load efficiency. (2) Motors designed to operate at a higher speed have higher efficiencies than those designed to operate at lower speeds. (3) Motors that have higher starting torques have lower full-load operating efficiencies than those with lower starting torques. (4) Three-phase motors have higher efficiencies than single-phase motors of the same horsepower rating.

Higher-Efficiency Motor Standards

The Energy Policy Act of 1992 (EPACT) addressed a wide range of energy issues, including motor efficiency. EPACT established minimum energy standards for the manufacture of a wide range of general purpose motors, those commonly found in facilities today. Starting in October of 1997, all general purpose T-frame, single speed, ploy-phase induction motors between one and 200 horsepower that are manufactured for sale in the United States must meet these standards. The standards, based on the nominal full-load efficiency of a motor, were an improvement by several percentage points over what manufacturers were offering as their standard-efficiency models. Figure 11.1 shows the difference by horsepower in full-load efficiencies between an average of what manufacturers were offering as standard-efficiency motors and the efficiency mandated by the standard.

While the standards have helped move manufacturers to developing and marketing higher efficiency motors, they do not address the large, installed base of electric motors that are operating in facilities today. There is no requirement to replace existing motors with high efficiency models. Even when these motors fail, they do not have to be replaced with higher efficiency models; the standard-efficiency motor can simply be rebuilt and placed back into service. Newly manufactured models must meet the higher-efficiency standards, but currently installed ones do not.

The savings produced by replacing a standard motor with a high-efficiency unit can be estimated from the following equation:

$$S = (0.746) \times (HP) \times (R) \times (N) \times (100/SE - 100/HE)$$

where: S = savings per year
HP = motor load in horsepower
R = electricity rate in $/kWh
N = annual run time in hours
SE = efficiency of the standard-efficiency motor
HE = efficiency of the high-efficiency motor

Figure 11.1
Performance of Standard and High-Efficiency Motors

Efficiency (%)

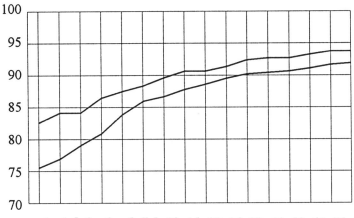

Horsepower

High -efficiency

Standard

Power Factor (%)

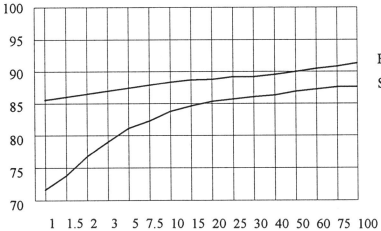

Horsepower

High-efficiency

Standard

Replacing existing motors with higher-efficiency models would save energy, but the size of the installed base and the cost of the replacement motors makes overall replacement cost prohibitive. There are applications and circumstances that make replacement an attractive option (see page 183 on motor replacement), but in general, replacement is not an option for maintenance managers. You must instead take steps to ensure that motor operating costs are minimized through effective maintenance. A comprehensive maintenance program that strives to improve motor operations through routine and preventive maintenance, proper sizing, and planned replacement will achieve savings comparable with a general motor replacement program at a fraction of the cost.

Building Motor Maintenance Programs Today

There is little doubt that good maintenance is the key to maximizing the performance of building electric motors. With good maintenance practices and routine diagnostic testing, nearly all motors can be expected to provide 10 to 15 years of continuous duty. Good maintenance practices will improve reliability, decrease energy requirements, and lengthen the life of the motor. Studies of organizations that have implemented comprehensive motor maintenance programs show that the programs produce savings ranging from 10 to 50 times their implementation cost in avoided motor failures and the resulting downtime.

In spite of the benefits, few facilities have implemented total maintenance programs. Most facilities are content to lubricate the bearings of motors on a regular basis and to visually check the operation of certain large or critical motors on a regular basis. Motor failures have become an accepted cost of operation. Even when motors repeatedly fail in the same system, it is generally attributed to a string of bad luck rather than poor maintenance practices.

The need to adequately maintain motors is readily apparent when one examines the causes of motor failures. Studies conducted by motor manufacturers and repair centers show that less than 5 percent of all motor failures can be attributed to old age. The causes of failure for the remaining motors is evenly split between mechanical and electrical factors.

The majority of mechanical failures involve worn or failed bearings. Bearing damage and failure are the direct consequence of factors resulting from the conditions under which the motor was operating: overly tightened belts, poor alignment between the motor and the driven load, too little lubrication, too much lubrication, use of the wrong lubricant, excessive vibrations, high operating temperatures, dirt, and moisture. All of these factors can be controlled or eliminated through sound maintenance practices.

Most electrical motor failures are the result of short circuits in the motor components where insulation has broken down. Factors that cause or contribute to the breakdown of the insulation include dust, moisture, overloading, high ambient temperatures, over voltage, under voltage, voltage surges, and voltage unbalances. As with the mechanical causes, most of the electrical factors can be controlled or eliminated through sound maintenance practices.

HOW TO DEVELOP A MOTOR MAINTENANCE PROGRAM ————————

Before examining the elements of a motor maintenance program, it is helpful to understand the basics of how electric motors operate. Understanding how they operate and what their basic components are will help in developing an understanding of the maintenance requirements for motors and what the energy implications are of those activities.

The vast majority of the motors used in facilities are induction motors. The induction motor is very popular because it is low in first cost, simple to operate, easy to control, inexpensive to maintain, durable, and available in a wide range of horsepower ratings. Few other motor types and configurations can compare with the induction motor.

The induction motor has two main parts: the stator and the rotor. The stator is the fixed component of the motor consisting of the primary windings, the laminated core, and the motor's supporting frame. The rotor consists of the rotating elements of the motor, including the secondary windings, the magnetic core, and the motor's shaft.

The induction motor is essentially a rotating transformer. When voltage is applied to the motor's stator windings, they generate a magnetic field in the stator. This field induces a secondary field in the rotor of the motor. The interaction of the two fields causes the rotor to turn, producing torque in the motor's shaft.

There are several different values used to rate the torque developed by motors. All torque ratings are for the motor operating at its rated voltage and power line frequency. The most commonly cited torque rating is full-load torque. The full-load torque rating is the torque developed by the motor when it is operating at its rated output and speed. The full-load torque rating is essential in matching motor characteristics to a particular driven load.

Another commonly used torque rating is the locked-rotor torque. Measured when the rotor is at rest, the locked-rotor torque is the minimum amount of torque developed by the motor when power is first applied during start-up. All driven loads present a certain amount of inertia to the motor during start-up. The locked-rotor torque of a motor must be high enough to overcome this inertia. Locked-rotor torque ratings are important to both performance and motor life. If the motor takes too long to overcome the inertia of the load, excessive heating in the motor's windings occurs, decreasing their performance and life.

Similar to locked-rotor torque, a motor's pull-up torque is the minimum torque developed by the motor while it is accelerating the load from rest to the rated operating speed. This torque must be sufficient to accelerate the load quickly enough to prevent overheating in the motor's windings.

Finally, there is breakdown torque. Breakdown torque is the maximum torque that a motor will develop without a sudden decrease in the motor's operating speed. Breakdown torque is typically between two and three times the full-load torque rating of the motor. This does not mean that the motor can be regularly operated at higher than full-load torque ratings. It simply indicates that the motor can still produce torque at this higher load rating. Operating at any point above the motor's full-load rating will increase the motor's operating temperatures, resulting in a decrease in motor life.

Rule of Thumb: When selecting a motor for a particular application, it is essential that the motor's characteristics, including operating speed, full-load torque, locked-rotor torque, pull-up torque, and breakdown torque, be matched to those of the load. Mismatches result in decreased motor life and increased energy use.

Step 1: Developing a Motor Inventory

One of the most important elements in a motor maintenance and energy conservation program is the development of a motor inventory. Before steps can be taken to reduce motor energy use and extend motor operating life, the maintenance department must know what is installed in the facility, where it is installed, what load it is driving, and if it is properly sized for the application.

Too often, maintenance managers have no idea of what is installed or where. As a result, motor maintenance is spotty. Motors that are highly visible may receive regular maintenance attention, while other, less visible motors may operate for years without any maintenance attention. Only when these units fail do they receive any attention.

Lack of an inventory of what is installed in the facility also contributes to a major cause of motor energy waste: oversizing. Unless there is a permanent record of what is installed, changes will take place over time. When a motor fails, the maintenance department is usually under pressure to get the system that it supports back in operation as quickly as possible. As a result, the failed motor may be replaced with one that does not exactly match the original motor's characteristics, simply because it may have been more readily available. In nearly all cases, this means that the replacement motor will have a larger horsepower rating than the one that it is replacing. The result is that the replacement motor will now be oversized for the application, and will be operating at only part load. Part load operation is the single biggest waste of energy in motor systems.

A complete inventory with a listing of all motors installed in the facility, as well as a history of their maintenance records, will ensure that motors receive the maintenance attention they require. Use Figure 11.2 to develop the motor inventory. Completing the inventory is a time-consuming but essential element in establishing the maintenance program. Information collected during the survey will provide the specific data needed to maintain the existing motors and reduce their energy requirements.

If the facility is too large and it would take too long to inventory all motors, break the survey down into at least two phases. The first phase should concentrate on two groups of motors: (1) those that are 5 horsepower and above; and (2) those that support essential operations, such as chilled or condenser water pumps. Once the first phase has been completed, additional motors can be added to the inventory as time allows.

Step 2: Record Motor Maintenance History

The maintenance history of a motor is particularly important in avoiding motor failures and improving motor energy efficiency. While historical operating data are valuable

Figure 11.2
Motor Inventory Data

1. Building: _____ 2. Room: _____

3. Load description: _____

4. Inventory number: _____ 5. Annual operating hours: _____

6. Date installed: _____ 7. Date rebuilt: _____

8. Special features: _____

Nameplate Data

9. Manufacturer: _____ 16. Model: _____

10. Serial number: _____ 17. Frame size: _____

11. Voltage: _____ 18. Phase: _____

12. Full-load amps: _____ 19. Horsepower rating: _____

13. NEMA torque class: _____ 20. Insulation class: _____

14. Service factor: _____ 21. Temp. rating: _____

15. Synchronous speed: _____ 22. Full-load rpm: _____

23. Comments: _____

_____

_____

_____

on most facility equipment, such data are a necessity when it comes to performing maintenance on electric motors. Electric motor failures seldom occur suddenly and without warning. In nearly all cases, they are the result of long-term operating conditions that lead to a slow deterioration of some component in the motor. Eventually the deterioration reaches the point where the motor fails. Keeping a maintenance history with the test results for the motor will show this trend of deterioration and will allow maintenance personnel to project when repair action is required, well before the motor actually fails.

In establishing a maintenance history for a motor, it is important to test a new or recently repaired motor soon after it is placed in service. This testing establishes a baseline for the motor's operation against which future motor tests can be compared. Sudden changes in motor testing results generally are an indication that a problem is developing. A slow change that takes years to develop is generally the result of normal wear and tear on the motor. Eventually, even these slow changes will create the need to overhaul a motor. By tracking the test data for several years, maintenance personnel will be able to project the best time to plan on overhauling or replacing the motor.

Use Figure 11.3 to record maintenance and test data from the motor. The interval between motor testings may be lengthened for those motors that show no rapid deterioration of motor electrical insulation or other components. Again, facilities that operate a large number of motors may find it necessary to break the development of baseline motor data down into two or more phases based on the size of the motor and the importance of the function that it supports.

Step 3: Determine Motor Size

The single most important factor in minimizing energy use by motors is for the maintenance program to properly match the motor capacity to the driven load. Induction motors operate very efficiently at or near their full-load horsepower rating. For example, a ten horsepower motor that meets the EPACT standard has a full-load efficiency of 90.2 percent. That efficiency remains fairly constant as the load is decreased until it reaches approximately 75 percent of the motor's full-load rating. Once the load has fallen below 75 percent of the motor's full-load rating, the motor's efficiency starts to decline more rapidly. By the time the load is reduced to 50 percent of the motor's full-load rating, the efficiency has fallen off to approximately 75 percent. Further decreases in load produce even greater declines in operating efficiency.

In addition to having lower operating efficiencies at reduced loads, the power factor of induction motors also declines. When an induction motor is operating at or near full-load, its power factor typically is close to 90 percent. As the load decreases, the power factor drops off fairly rapidly. At 50 percent load, the power factor for most induction motors is approximately 60 percent. Figure 11.4 shows the relationship between motor loading and efficiency and power factor.

Low power factor resulting from underloaded induction motors may have a direct impact on a facility's electrical energy costs. Many utility companies impose a power

Figure 11.3
Motor Test Data

1. Building: _____ 2. Room: _____

3. Load description: _____

4. Inventory number: _____ 5. Date of test: _____

6. Ambient temperature: _____

Test Data

6. Voltage by phase: _____ _____ _____

7. Current by phase: _____ _____ _____

8. Input kW: _____ 9. Operating speed: _____

10. Bearing temperature: _____ 11. Surface temperature: _____

12. Vibration amplitude: _____

13. Megohmmeter resistance: _____

14. Polarization index test: _____

15. Hi-pot test: _____

16. Surge comparison test: _____

17. Motor shaft endplay: _____

18. Comments: _____

_____

_____

_____

Figure 11.4
The Effect of Part-Load Operation
on Motor Efficiency and Power Factor

Percent Load	Efficiency (%)	Power Factor (%)
10	30	25
20	60	40
30	68	45
40	75	53
50	80	61
60	83	68
70	84	71
80	85	75
90	85	78
100	85	80
110	84	81
120	83	81

factor penalty clause in their rate schedule when a facility's power factor falls below a set value, typically 0.85 or 0.90. Other utility companies bill their demand charges based on the facility's peak kVa requirement, a measure that already factors in the effects of power factor. In both cases, operating with a low system power factor as the result of underloaded induction motor operation will increase energy costs.

Oversized motors are a common problem in facilities. Surveys of facilities show that nearly one-third of the motors installed are operating at 50 percent or less of their rated load. This means that one-third of even the high-efficiency motors are actually operating at less than 60 percent efficiency. These figures do not even take into consideration motors that drive variable loads that fluctuate widely. Properly sizing the motors in many of these applications will produce sufficient savings to justify replacing the existing motor even though it may be in good operating condition.

There are a number of factors that contribute to the oversizing of motors. Systems designers typically build in a certain amount of oversizing, typically 10 percent, to allow a safety factor for unforeseen system conditions. Actual operating conditions for induction motor fans and pumps may be different that what was anticipated during design, resulting in oversizing of the motor. Changes may have been made in the system over the years that have lowered the load on the motor. A replacement motor with a higher horsepower rating may have been installed at some time in the past when the original motor failed. The result is that most induction motors are oversized for the application. Many are greatly oversized.

The survey of the installed motor will identify applications where the existing motor is oversized for the load it is driving. Although the savings produced by replacing oversized motors with ones properly sized for the load are significant, few facilities can afford to replace one-third or more of their existing motors, particularly if the existing motors are in good operating condition. While the added savings produced by replacing an oversized, standard-efficiency motor with a properly sized, high-efficiency motor will make the replacement program even more cost-effective, replacement still will be too expensive for nearly all facilities.

There are two alternatives: swapping existing motors, and scheduled replacement. The first, swapping motors, uses the information gathered by the motor inventory to identify locations where motors that are oversized for a particular application would be better matched to the load in another application. Swapping is difficult to implement as the characteristics of the motor, including motor frame size and torque ratings, must match the application requirements where they are being installed. Swapping is most effective in facilities with a large number of motors of various sizes.

The second alternative is scheduled replacement. The maintenance record-keeping program will identify motors that are developing problems and are at risk of failing. Replacement of the oversized motors that are in risk of failure with properly sized, high-efficiency models before they actually fail will reduce energy costs and prevent unscheduled interruptions of service. The best candidates are those motors that are indicating developing problems, have been repaired a number of times in the past, have annual hours of operation in excess of 2,000, and are loaded to less than 50 percent of their rated capacity.

ELEMENTS OF THE MOTOR MAINTENANCE PROGRAM

The purpose of the maintenance program is to implement activities that will help ensure that the maximum service life of the motor is achieved, and that the motor performs at a minimum energy cost. To accomplish this, maintenance activities will have to be performed on the electrical system powering the motor, the motor itself, and the connection to the load being powered by the motor. The frequency with which the activities are performed varies with factors related to the application, including the annual operating hours of the motor, the environment in which the motor operates, the type of load connected to the motor, how often the motor is started, and the critical nature of the operation supported by the motor.

Figure 11.5 lists the recommended maintenance activities for general purpose motors operating under normal conditions. Use the figure to identify the maintenance tasks to be performed. Use of the figure should be coordinated with test data recorded in Figure 11.3

System Voltage Testing

Efficient motor operation requires that the electrical system powering the motor supply provide electrical service under very specific operating conditions, including proper voltage, balanced voltages across phases, and a minimum of harmonic voltages.

Operating a motor at a line voltage other than its design voltage changes the motor's performance characteristics and reduces its operating life. Too high a voltage results in saturation of the motor's magnetic core, increasing iron losses. Too low a voltage results in reduced output torque and increased copper losses. To ensure that the motor is operating at its rated voltage, the voltage should be checked at the motor while it is operating under full-load. Any variation beyond ± 5 percent of the motor's rated voltage should be investigated and corrected.

Three-phase motors have the additional voltage requirement that the voltage supplied by each of the system's phases must be the same or balanced. Any voltage variation between the phases decreases the torque developed and increases heat generated within the motor. Even small imbalances between the phases result in large increases in motor losses. Take voltage measurements across each phase of the power system at the motor with the motor operating under full-load. Any voltage variation between phases greater than 1 percent should be investigated and corrected.

Electronic equipment, such as power inverters and variable frequency drives, generates harmonic voltages that then are present throughout the power system. When these harmonic voltages are present in the service powering motors, they induce a harmonic current in the motor's coils that produces a magnetic field within the motor. This harmonic field increases copper and iron losses, and raises the operating temperature of the motor. ***Rule of Thumb:*** The facility's power distribution system should

Figure 11.5
Motor Maintenance Activities

1. Building: _____ 2. Room: _____

3. Load description: _____

4. Motor inventory number: _____

5. Date: _____

Maintenance Tasks

Clean motor and fan	☐	_____
Check motor-load alignment	☐	_____
Check belt tension	☐	_____
Lubricate (if required)	☐	_____
Inspect coupling/pulley	☐	_____
Check mounting bolts	☐	_____
Inspect motor base	☐	_____
Inspect electrical contacts	☐	_____

be periodically checked for the presence of harmonic voltages. If the harmonic voltages are high enough, filters will have to be installed on the equipment generating them.

Routine Inspection

The easiest and one of the most important maintenance tasks that can be performed is a routine inspection of the motor. Inspect the motor for cleanliness on a regular basis. It is important that motors be kept as clean as possible. Dirt on the motor case impedes heat transfer from the motor housing, causing the motor to run at a higher temperature and decreasing motor life. For example, a ten-degree centigrade increase in operating temperature will reduce the life of the insulation on the motor windings by 50 percent.

Dirt can also be blown into the interior components of the motor by the cooling fan. Sand, dirt particles, metal shavings, and other debris form grit that abrades the insulation on the motor windings and increases wear in the bearings. Dirt also can clog the cooling fan air intake, increasing the operating temperature of the motor.

Rule of Thumb: Once a year, the electrical equipment supporting the motor should be inspected and cleaned. More frequent cleanings are required in dirty environments. All contacts in relays, circuit breakers, and fuses should be checked for pitting and discoloration. If a thermal scanner is available for use, scan all electrical contacts while the equipment is operating to identify hot spots that may be an indication of a connector that is loose or a contact that is corroded.

Lubrication

Lubrication, a critical factor in motor bearing and winding life, is a balancing act. Too little lubrication increases friction losses and leads to early bearing failures. Too much lubrication also increases friction losses and can force grease out of the bearing housing and onto the motor windings where it attacks the insulation and can lead to short circuits in the windings.

Rule of Thumb: There is no set interval for when a motor should be lubricated. A large motor subject to frequent starts, vibrations, or high bearing loads may require lubrication every two or three months. In contrast, a motor that is used intermittently, is protected from harsh environmental conditions, and has a properly aligned connection to the driven load, may not require lubrication more often than once every two or three years.

The best way to ensure proper lubrication of a motor is to follow manufacturer's recommendations, not only for frequency of application, but also for the type of lubricant best suited for the motor. Mixing of incompatible lubricants may cause them to harden or separate, resulting in early bearing failure.

Insulation Testing

One of the most common problems encountered in motors is degradation of the motor windings. Heat, dirt, moisture, and other contaminants act to break down the insulation on the windings This breakdown is gradual and can be detected through routine testing. If it continues undetected and uncorrected, the motor will eventually fail. Fortunately, most developing faults in motor insulation are readily detected through a number of tests.

The more important aspect concerning the results of insulation tests is the trend that the readings show over time. If readings are very close to one another over a period of years, then the insulation is stable. If the readings show a slow decline over time, then the insulation is slowly deteriorating. The rate of deterioration will determine how quickly action will have to be taken. In contrast, a sudden change in one or more readings indicates that something has happened and the motor may be at risk. For these reasons, it is very important that all readings be recorded and kept on file for regular review.

The insulation resistance test, also known as the megohmmeter test, is the simplest and most common test performed to evaluate the condition of the motor insulation. Megohmmeter tests identify turn-to-turn, coil-to-coil, and phase-to-phase faults within the motor. A high reading indicates good, dry windings, and while a low reading may indicate failing insulation, it may also be the result of moisture in the motor's windings. Follow the manufacturer's recommendations as to what is an acceptable value.

The dielectric absorption or polarization index test is nearly identical to the megohmmeter test that evaluates the condition of the insulation on the motor's windings. While it takes longer to perform, it produces more accurate results.

Another test of the condition of motor insulation is the DC high potential or hi-pot test. During the testing, a high voltage is applied to the motor's windings in a series of steps. As the voltage is increased, small weaknesses in the insulation can be identified—weaknesses that may lead to motor failure. Although the hi-pot test is very effective in identifying weaknesses, it can cause failure in an operating motor that has weak but not yet failed insulation. For that reason, most manufacturers recommend using the hi-pot test sparingly. Hi-pot tests identify all potential faults within a motor.

The surge comparison test is useful in identifying ground faults within a motor, but cannot detect turn-to-turn, coil-to-coil, or phase-to-phase faults.

It is recommended that all motors in the testing program be regularly tested using the megohmmeter and the surge comparison tests, with an occasional hi-pot test. ***Rule of Thumb:*** Testing intervals will vary with the application, but for most facilities the recommended intervals are every three months for motors of 25 horsepower or more, every six months for motors of 10 to 25 horsepower, every six months for motors of less than 10 horsepower that serve critical loads, and every year for all other motors. Less frequent testing will be required for motors that operate less than 2,000 hours per year.

Alignment

Proper alignment of the motor and driven load is important for long motor bearing and coupling life. Even slight misalignments increase the lateral loads on bearings and couplings, causing them to run at higher-than-normal temperatures and to wear excessively. Proper alignment is also important in belt-driven equipment.

Proper alignment means that the centerline of the motor shaft and the centerline of the load shaft coincide. The shafts must be parallel and must have no angular misalignment. Most alignment problems originate from improper installation, excessive vibrations, or broken mounts.

The alignment of all motors should be checked shortly after the motor is initially installed or after an overhaul of the motor. The alignment should be checked again if unusual conditions develop, including the detection of vibrations, abnormally high bearing temperatures, unusual noises, or signs that the motor or load have been bumped.

Rule of Thumb: At least once a year, the motor mounting structure should be examined for cracks, broken or loose bolts, or any other conditions that would allow the motor to shift its position relative to the load. If problems are found, they should be corrected and the motor load alignment checked.

Vibration Measurement

Excessive vibrations damage motor bearings and can crack the insulation on motor windings, leading to winding failure. There is a large range of causes for vibrations. Assuming that the motor was vibration free when it was first installed, vibrations will increase over time as the result of misalignment of the motor and load, a bent motor shaft, loose motor and load mounting bolts, accumulation of dirt on rotating elements within the motor, or worn bearings.

Motors should be tested regularly for vibrations and the readings recorded for future reference. The testing interval depends on the motor size and the critical nature of the load. *Rule of Thumb:* It is recommended that vibration measurements be taken every three months for motors of 25 horsepower or more, every six months for motors of 10 to 25 horsepower, every six months for motors of less than 10 horsepower that serve critical loads, and every year for all other motors.

A gradual increase in the level of vibrations over the life of the motor is considered to be normal. However, if the readings show a sudden, sharp increase, the cause should be investigated before it leads to a possible motor failure and unscheduled load shutdown.

IMPLEMENTING THE MOTOR MAINTENANCE PROGRAM

If motors have been ignored for any period of time, the motor maintenance program will have to be phased in. Facilities simply have too many motors to be able to implement the entire program all at once. Data will have to be collected on what motors are installed where, and the functions that they support. Maintenance histories will have to be reconstructed from work orders, motor repair logs, and purchasing records. Testing and other maintenance activities will have to be carried out. While these activities are being carried out, motors will still be failing due to the lack of ongoing maintenance in the past.

To obtain the greatest benefits from the program in energy savings, reduced maintenance costs, and improved reliability, concentrate the first phase of the program on the largest motors, motors that have annual run-times in excess of 2,000 hours, and motors that support critical functions. Even then, most facilities have a large number of motors that fall into these three categories. If the number is too large for implementation all at once, further split the motors into two or more groups based on the priorities for the facility. If the major goal is energy management, start with the larger motors with long run-times. These are the motors that will offer the greatest potential for energy savings due to oversizing and possible replacement with high-efficiency motors. If reliability is the primary concern, start with the motors that support critical systems, particularly those that do not have a backup motor installed.

Once the initial phase has been implemented, expand the program based on the priorities of the facility. It will take several years for the program to take full effect, particularly if motor maintenance has been ignored in the past.

Motor Replacement Program

Even with a well-implemented maintenance program, motors will fail. When they do, you will have the option of rebuilding the existing motor or replacing it with a new one. In most facilities, the repair or replace decision is based primarily on first costs and speed. Repairing an existing motor is almost always less expensive than replacing it. In addition, the existing motor has failed, and there is a need to get the system back up and running as quickly as possible. But speed and first costs are only two of the factors that must be considered. Motor repair or replacement decisions must also take into consideration energy and maintenance factors.

For example, nearly all motors that are five years old or older are standard-efficiency models. Older motors have even lower operating efficiencies. All new motors being sold today must meet the new energy efficiency standards and are between 2 and 6 percent more energy efficient than older standard-efficiency motors. Replacing any existing failed motor with a higher-efficiency model will automatically reduce energy requirements. The question is, will it reduce energy requirements enough to justify the cost differential between repair and replacement.

Depending on the cause of the motor failure, even when it is fully rebuilt, the motor may not achieve the same level of operating efficiency it had before the failure. Heat generated during the burnout of motor windings can degrade laminations, distort the stator core, and alter the air gap, all of which decrease the efficiency of the motor.

The time to make the repair or replacement decision is well before the motor has failed. Go through the maintenance inventory and history for the motors installed in the facility. Start with the larger and more critical motors and examine a number of factors related to their operation. How old is the existing motor? What is the condition of the existing motor? Is the performance of the motor declining? Is it properly sized for the application or is it oversized? What is the full-load efficiency of the existing motor? How many hours does it operate each year? If the existing motor were replaced with a high-efficiency model, how long would it take to recover the installation costs? Developing the answers to these questions in advance will help in identifying the best course of action when the motor fails.

The ideal candidate for replacement is an older motor that operates 8,760 hours per year. Depending on its operating efficiency, these motors typically use between 10 and 20 times their purchase cost in electricity every year. The typical energy efficient motor uses 8 to 12 times its purchase price in electricity each year. In applications where the motor operates 8,760 hours per year, the cost differential between repair and replacement is recovered in energy savings in less than one year. Not all motors operate year round so it will be necessary to carry out the payback calculations based on actual operating conditions. ***Rule of Thumb:*** If repair costs total more than 50 percent of the replacement costs, replace the motor.

A word of caution when high-efficiency motors are installed as replacement units. One of the characteristics of high-efficiency motors is their lower slip rates in comparison to standard motors. As a result, when a high-efficiency motor is installed as a replacement for a standard motor, it will operate at a slightly higher speed. Motor load is very sensitive to motor speed, increasing with the cube of the speed. Even a small increase in motor speed will result in a large increase in the motor's energy requirements, enough to negate the benefits of a high-efficiency motor. Therefore, it may be necessary to modify the motor drive to ensure that the load is not driven at a higher speed than what it was originally designed for.

The Impact of Cycling on Motor Life

The large electrical load that motors represent has made them a target for energy management programs. There are two general approaches that energy managers have taken to shutting down motors in order to reduce energy costs: unoccupied hours shutdown, and shutdown for demand limiting.

Unoccupied hours shutdown typically lasts for several hours. The shutdowns save not only the energy that would have been used by the motor, but there are also secondary energy savings that result from shutting down the system that the motor supports. For example, shutting off a motor driving an air handler saves the fan motor energy, the heating and cooling energy that would have been supplied by the system, and the energy that would have been used to heat or cool ventilation air.

Shutdowns for electrical demand limiting purposes are different. Most facilities operate with a 15 or 30 minute demand period. The peak demand for the facility typically occurs sometime between 10:00 A.M. and 4:00 P.M. During this period, demand limiting equipment shuts off large electrical loads, including motors, for short periods of time in an attempt to limit the facility's demand for electricity. Each 15 or 30 minute period represents another window when equipment may have to be shut down. While the loads that are temporarily shut off will vary from one demand window to another, the net effect is that motors may be stopped and started frequently during the day.

More frequent starts decrease motor life. Each time a motor is started, its components are subjected to stresses—stresses that accumulate over time. The main concern with frequent starts is heat. Every time a motor is started, it is subjected to a locked rotor current that is nearly six times the normal full-load operating current for the motor. This current drops off gradually as the motor accelerates, but it results in a large buildup of heat within the motor. This heat buildup is particularly large for high inertial loads, such as building fans and pumps.

The heat that is generated during motor start-up raises the temperature of the motor's windings, accelerating the deterioration of the insulation. The more often the motor is started, the more rapid the rate of deterioration, particularly if the time between starts is short. While all starts contribute to the deterioration of the insulation on the motor's windings, it is the frequent starts over a short period of time that are most damaging.

Manufacturers take into consideration the fact that a motor will be subjected to stresses during start-up. NEMA standards require that a motor at ambient temperature be able to complete two starts from rest without exceeding its rated load operating temperature. Similarly, NEMA standards require that a motor operating under full-load be capable of restarting once without exceeding its rated operating temperature. Too frequent starts over a short period of time will easily exceed these conditions and will result in overheating within the motor's windings.

If motors are cycled to help limit the electrical demand for a facility, use the following rules of thumb to limit the detrimental effects of frequent motor starts on the life of the motor: For motors less than 5 horsepower, the minimum off time should be five minutes. For motors between 5 and 15 horsepower, the minimum off time should equal one minute per horsepower. For larger motors, consult the motor manufacturer for suggested minimum off times.

If frequent starts are a requirement, one option available is the installation of a reduced voltage starter. Reduced voltage starters allow the motor to start more slowly by limiting the starting current to a value well below normal. This reduced current will limit the heating effects on the motor's windings. Reduced voltage starters also limit the wear and shock damage to the motor-to-load drive train and the load itself. Solid state reduced voltage starters allow maintenance personnel to adjust the starting characteristics of the motor to match the requirements of the particular load connected. In addition to benefiting motors that are frequently cycled, reduced voltage starters will help prolong the life of motors driving large fans, compressors, and positive displacement pumps.

SUMMARY

A planned maintenance program for motors is one that offers a high rate of return through reduced energy costs, motor losses, and downtime. Compared to facilities that perform only the most basic motor maintenance, a facility that implements a comprehensive maintenance program can expect to reduce motor failures by an average of

90 percent, reduce the cost of maintaining the motors by more than 50 percent, and reduce the energy requirements of the motors by 20 to 30 percent.

The basic elements of the program include an inventory of the motors installed in the facility, the compilation of their maintenance history, an evaluation of the suitability of each motor to its application, development of a scheduled maintenance program that includes testing of the motors, and the establishment of a motor replacement policy that addresses motor repair considerations. Although the programs take time and effort to establish, the savings achieved through reduced motor failures and system downtime often pay for their implementation in the first year.

Section 4

How to Reduce the Impact
of Aging on the Energy Efficiency
of Components
of the Building Envelope

Chapter 12

Doors and Windows

This chapter presents guidelines and procedures on the maintenance of building exterior doors and windows. The information presented in this chapter is directed toward the types of doors and windows most commonly found in commercial and institutional facilities. The maintenance activities identified in this chapter are intended to be used in establishing a scheduled maintenance program for exterior doors and windows. By following the maintenance activities identified in this chapter, you will be able to implement a door and window maintenance program that will enhance the appearance of your facility, extend the operating life of these building components, improve their performance, reduce breakdown maintenance costs, and reduce facility energy requirements.

DOOR AND WINDOW ENERGY MAINTENANCE CONSIDERATIONS

Doors and windows play a very large role in the appearance, accessibility, security, and energy efficiency of commercial and institutional facilities. They serve as the front line defense against the elements and intruders. They are expected to keep out the cold, the heat, the rain, the snow, the wind, and the unauthorized, while providing a view and a means of easy egress for occupants and visitors. They must function 24 hours a day, 365 days a year, exposed to temperature swings of more than 100 degrees, winds, ultraviolet radiation, rain and snow, acid rain, and a whole host of other atmospheric pollutants. And they must perform these functions while retaining their appearance.

Doors and windows can be thought of as holes in an otherwise solid building exterior. Their energy loss rates per square foot, even for sound, well-maintained units, average four or five times the rate for typical wall and ceiling construction. For worn, poorly maintained, or improperly closing units, the rate of energy loss is even greater. For example, a loose fitting door that is not weather-stripped allows nearly the same amount of air to infiltrate into a facility as does a 12 inch by 12 inch hole in the wall. The rate is even higher for doors that do not fully close.

Doors and windows also are a significant investment for the facility in terms of first costs and ongoing maintenance. Maintenance costs on a square foot basis exceed those for all other building envelope components, with the exception of some roofs.

Having to replace existing doors and windows is also a major disruption to the building occupants. Entrances must be closed and pedestrian traffic rerouted. Furnishings have to be moved. Occupants may have to be temporarily relocated. Interior finishes will have to be repaired or replaced.

Complicating the performance expectations for doors and windows is the fact that they are heavily used, and abused. No other component in the building envelope is subjected to the daily wear and tear that doors and windows are. Both are high-maintenance items that require constant attention if they are to function properly.

Doors and windows are routinely exposed to a number of different forces that act to decrease their performance, energy efficiency, and operating life. They are subject to accidental damage from users and deliberate damage from vandals. Vibrations originating from both within and outside the facility can loosen hardware, structural components, and sealants. Water can cause damage in the form of corrosion, pealing finishes, and rot. Water vapor can condense within the units, resulting in decreased insulating properties and increased rates of corrosion and rot. Exposure to ultraviolet light and atmospheric pollutants can change material properties that accelerate the aging of door and window components. Thermal movement can overstress sealants resulting in water penetration. Freeze/thaw cycles, particularly if water has gained access to the interior of door and window components, can expand hairline cracks into major gaps, admitting even more water and air. Protecting against these forces requires implementation of a well-planned maintenance program.

In spite of this high required level of attention, few organizations have a planned approach to door and window maintenance. Surveys of these facilities find that the majority of the exterior doors have maintenance problems sufficiently significant to interfere with their operation. There are doors that hang open; locks and latches that are difficult to operate; windows that have large air gaps between the sash and frame. There is missing weather-stripping; deterioration of frame members. All are common problems found in facilities that do not have a planned maintenance program for doors and windows.

If doors and windows are to be able to perform anywhere close to expectations, they must be properly maintained through planned maintenance, not breakdown maintenance. Breakdown maintenance is reactive maintenance. The maintenance department waits until something fails to operate and is called to their attention by the building occupants; then they react. The problem with this approach to maintaining doors and windows is that the type of failures that prompt maintenance calls are only the most extreme, such as doors that cannot be closed or opened, doors that cannot be locked, windows that have broken glass, or windows that are stuck open or closed. Other problems that impact the performance of the doors and windows but do not prevent their use simply are not reported. Bad seals, missing weather-stripping, doors and windows that do not fully close—all are examples of the types of maintenance problems that are seldom reported yet cost the facility in terms of energy use and shorten door and window life. Under breakdown maintenance, if something is not reported as being broken, it will not get fixed.

One exception to this breakdown approach to maintenance is exterior painting. Many maintenance organizations have implemented a scheduled program for painting the exterior surfaces of wood window. These programs typically establish a five-to-seven-year painting cycle for building exteriors, including wooden doors and windows. Although the programs are driven more by appearance concerns than by maintenance factors, they have proven effective in enhancing the appearance of facilities while helping to preserve exterior wood trim and components, such as doors and windows.

Door Construction

There is a wide variety of types of doors installed in facilities, including swinging, bypass sliding, surface sliding, pocket sliding, revolving, and storefront entrance. They can be manually or electrically operated. Nearly all are automatically closing.

There also is a wide variety of types of construction of exterior doors, including wood flush, wood rail and stile, hollow metal, and storefront. While specific maintenance requirements vary with the type of materials used to construct the door, all door types require regular inspections and maintenance if they are to perform adequately.

Commercial grade wood doors, including both flush and rail and stile construction, can be installed in wood or metal frames. Their popularity is due in part to their appearance and durability. Their greatest limitation is their susceptibility to damage from water and moisture, particularly in applications where they are not protected by overhangs. Uncorrected, it will eventually lead to rot and decay.

Hollow metal doors consist of a painted metal outer layer over a honeycomb or polystyrene core. To resist corrosion, the metal is generally hot-dipped galvanized, protecting the metal on both sides. Metal doors are popular due to their sturdy construction and resistance to physical damage. Their most serious limitation is corrosion. Even with a protective layer of galvanizing, damage to the door's finish can extend through the protective layer, leading to corrosion.

Storefront doors consist of a two-to-four-inch metal stile and rale frame and tempered glass. The metal frame is generally made of anodized aluminum to resist scratching and corrosion. Storefront doors are popular due to the full view they give people entering and leaving a facility. Their high rate of use subjects them to high levels of abuse and wear.

Door Hardware

One of the most critical elements in proper door operation is the hardware. Door hardware, like the door itself, is subject to wear and abuse every time the door is opened. Metal components corrode. Kick and push plates can fall off, exposing the door surface to wear and abuse. Thresholds fill with dirt and debris, causing the door to hang open. Hinges wear and loosen, allowing the door to sag. Closers won't allow the door to fully close, or they allow the door to shut too quickly. Weather-stripping fails, increasing the rate of air infiltration into the facility. Of these problems, the most common involve hinges, closers, and weather-stripping.

The most common types of hinges used on exterior doors are butt, surface, and invisible. Butt hinges, used with both wood and metal doors, are mortised into the door edge and jamb. Surface hinges are similar to butt hinges, but are not mortised into the door edge or jamb. Invisible hinges are concealed in a mortised hole in the door head and jamb. All three hinge types require periodic lubrication to minimize wear on the hinge pin, and regular inspection of the mounting screws to ensure that they are tight. Loose hinge mounting screws are generally a sign of improper alignment of the door.

Proper door closer operation is very important for the life of the door and for energy conservation purposes. A door that closes too quickly causes wear and impact damage to door and door hardware components. A door that fails to fully close results in large amounts of air infiltration. There are two basic types of door closers used, hydraulic and pneumatic. For commercial applications, the hydraulic closer is more common. Hydraulic closers contain a cylinder, an oil reservoir, and a flow adjuster to regulate the closing speed of the door.

Door closers must be tested and adjusted on a regular basis, the frequency depending on the level of use for the door. Properly adjusted closers will control the door's operation throughout its opening and closing swing.

There are several types of weather-stripping used on exterior doors. The most common ones are metal spring strip, interlocking metal strip, vinyl gasket, woven pile, and plastic. All are prone to wear as the result of normal door operation, and to abuse from impact as people and objects pass through the door. Common damage to weather-stripping includes crushing, displacement, and tearing. Crushed or torn weather-stripping fails to seal the gaps between the door and its frame, increasing the rate of air infiltration. Displaced weather-stripping also fails to seal the gap, or it may prevent the door from properly closing. All weather-stripping must be closely inspected on a regular basis for damage.

Window Construction

There is a wide variety of different styles of window construction used in facilities today. Window types include fixed, casement, awning, sliding, double hung, and jalousie. For these window types, there are three major factors that vary: operability, frame construction, and the glazings. Maintenance requirements vary with the materials used and the type of construction.

Operability, more than any other factor, determines the level of maintenance required for windows. In nearly all applications, fixed windows require far less maintenance than operable ones. Even windows that are designed to be opened only for cleaning have lower maintenance requirements than operable windows. Fixed window units are less subject to wear and have far fewer moving parts. Gaskets and sealants are subjected to less stress and in general perform better, providing lower rates of air infiltration.

There are three major types of frame construction used in windows: wood, metal, and vinyl. Each type offers specific advantages and disadvantages and has its own maintenance requirements.

1. *Wood framed windows,* constructed from Ponderosa pine, western pine, and spruce, are widely used in facilities today. The popularity of wood is due in part to its thermal stability, good heat transfer properties, and its ability to be milled in a wide variety of shapes and sizes for nearly any window requirement. Its biggest limitation is its susceptibility to damage from water and moisture. To help resist the effects of exposure to water and to eliminate the ongoing need for painting, most window manufacturers supply their products with vinyl or aluminum cladding. Cladding reduces the exterior maintenance requirement to annual inspections and periodic recaulking.

2. *Metal framed windows,* constructed from either aluminum or steel, are an alternative to wood framed units where windows are subjected to high structural or wind loads, or where the facility has stringent security requirements. Metal framed windows have higher heat losses than wood ones. To reduce the rate of heat loss, manufacturers build a thermal break made of polyurethane or neoprene into the frame to separate interior metal surfaces from exterior ones, greatly increasing the thermal resistance of the frame.

Metal frames come with either an anodized or painted protective coating. Both types of coatings are designed to last the life of the window and require minimal maintenance. They should, however, be inspected regularly for corrosion as the finish can be damaged by scratching or impact.

3. *Vinyl framed windows,* once restricted to the residential window market, have slowly been making their way into the commercial and institutional marketplace. To improve the strength of the frames in applications having larger windows, manufacturers frequently build their frame around an extruded aluminum core. Constructed from PVC, vinyl window frames offer good thermal properties, long lives, and very low maintenance requirements.

Window glazings come in a wide variety of materials, thicknesses, and finishes. They can be single, double, or triple glazed. The glazings can be independently mounted and gasketed, or they can be combined in a sealed unit that is filled with an inert gas. The glazings can be clear, tinted, reflective, heat absorbing, or coated with low emissivity coatings. Each has its advantages and disadvantages for various applications depending on the needs of the facility. Maintenance requirements are limited primarily to cleaning and inspecting for damage.

THREE WAYS DOORS AND WINDOWS CONTRIBUTE TO THE FACILITY'S ENERGY USE

There are three major ways in which doors and windows contribute to the facility's energy use: conduction of heat into or out of the facility, transmission of radiant energy into or out of the facility, and infiltration of outside air into the facility. High-quality door and window designs strive to limit energy losses. One goal of door and window maintenance programs is to see that those losses remain low over the life of the unit.

The rate at which heat is lost or gained through doors and windows by the process of conduction depends on the overall thermal conductivity of all of the door or window components. The most common measurement for thermal conductivity is the U-value. Calculated in Btu per square foot per degree Fahrenheit temperature difference across the door or window, the U-value is a measure of the rate at which heat is conducted through the door or window. The lower a unit's U-value, the higher its energy efficiency with regard to conduction. Figure 12.1 lists U-values for common wood and steel doors without glazing. Figure 12.2 lists U-values for common types of windows and glazed doors.

Although the procedure used to calculate the U-value for a particular door or window is straightforward, it should be considered to be only approximate. The actual U-value depends on a number of conditions particular to the installation. For example, a significant factor contributing to the door's or window's U-value is the resistance of the air film at the interior and exterior surfaces of the unit. Any movement of air across those surfaces, such as from wind or from a HVAC system discharge, will change the actual U-value.

The second way in which doors and windows contribute to the facility's energy use is through the transmission of radiant energy into or out of the facility. Sunlight, consisting of primarily heat and visible light, enters through the door and window glazings, increasing the cooling load, while radiant energy is lost from the building interior at night, increasing the heating load. Depending on the particular glazing used, the amount of solar radiation passing through the doors and windows ranges between 10 and 90 percent. With special low emissivity coatings, radiant energy losses through the glazings at night can be controlled.

The solar transmission characteristics of window and door glazings is measured by their shading coefficient. The lower the shading coefficient, the less solar energy is transmitted by the glazing. For example, a single layer of one-eighth-inch clear glass has a shading coefficient of 1.0. Two layers of one-quarter-inch clear glass has a shading coefficient of 0.80. Figure 12.3 lists common door and window glazing materials and their shading coefficients.

How much solar energy passes through the door and window glazing directly influences the energy use of the facility. During the cooling season, all solar energy entering the facility ends up as heat that must be removed by the facility's air conditioning system. During the heating season, the solar energy can help to reduce the heating load but it can also create areas that are warm enough to require the use of mechanical cooling.

The glazing installed in the doors and windows also influences energy used for lighting in these areas. Since the shading coefficient of the glazing limits both visible light and infrared energy passing through the glazing, high shading coefficients result in the need to operate additional lighting systems in these areas.

The third way in which doors and windows contribute to the facility's energy use is

Figure 12.1
U-Values for Wood and Steel Doors Without Glazing

Nominal Thickness (in.)	Wood Doors	U-Value
1 3/8	Panel door, 7/16 inch panels	0.57
1 3/8	Hollow core flush door	0.47
1 3/8	Solid core flush door	0.39
1 3/4	Panel door with 7/16 inch panels	0.54
1 3/4	Hollow core flush door	0.46
1 3/4	Panel door with 1 1/8 in panels	0.39
1 3/4	Solid core flush door	0.40
2 1/4	Solid core flush door	0.27
	Metal Doors	
1 3/4	Fiberglass or mineral wool core with steel stiffeners, no thermal break	0.60
1 3/4	Paper honeycomb core without thermal break	0.56
1 3/4	Solid urethane foam core without thermal break	0.40
1 3/4	Solid fire rated mineral fiberboard core without thermal break	0.38
1 3/4	Polystyrene core without thermal break	0.35
1 3/4	Polyurethane core without thermal break	0.29
1 3/4	Polyurethane core with thermal break and wood perimeter	0.20
1 3/4	Solid urethane foam core with thermal break	0.20

Note: Reprinted with permission from *1997 ASHRAE Handbook, Fundamentals.*

Figure 12.2
U-Values for Door and Window Glazing

Glazing Type	U-Value
Single Glazing	
1/8 in. glass	1.04
1/4 in. acrylic/polycarb	0.88
1/8 in. acrylic/polycarb	0.96
Double Glazing	
1/4 in. airspace	0.55
1/2 in. airspace	0.48
1/4 in. argon space	0.51
1/2 in. argon space	0.45
Double Glazing, e = 0.60 on surface 2 or 3	
1/4 in. airspace	0.52
1/2 in. airspace	0.44
1/4 in. argon space	0.47
1/2 in. argon space	0.41
Double Glazing, e = 0.40 on surface 2 or 3	
1/4 in. airspace	0.49
1/2 in. airspace	0.40
1/4 in. argon space	0.43
1/2 in. argon space	0.36
Double Glazing, e = 0.20 on surface 2 or 3	
1/4 in. airspace	0.45
1/2 in. airspace	0.35
1/4 in. argon space	0.38
1/2 in. argon space	0.30
Double Glazing, e = 0.10 on surface 2 or 3	
1/4 in. airspace	0.42
1/2 in. airspace	0.32
1/4 in. argon space	0.35
1/2 in. argon space	0.27
Triple Glazing, e = 0.20 on surface 2,3,4, or 5	
1/4 in. airspace	0.33
1/2 in. airspace	0.25
1/4 in. argon space	0.28
1/2 in. argon space	0.22

Note: Reprinted with permission from *1997 ASHRAE Handbook, Fundamentals.*

Figure 12.3
Typical Door and Window Glazing Coefficients

Type of Glazing	Shading Coefficient	Visible Light Transmission (%)
Single Glazed		
Clear	1.00	90
Gray tint	0.69	43
Bronze tint	0.71	52
Green tint	0.71	75
Reflective	0.51	27
Single Glazed, low-e		
Clear	0.74	84
Gray tint	0.50	41
Bronze tint	0.52	49
Green tint	0.56	71
Double Glazed		
Clear	0.84	80
Gray tint	0.56	39
Bronze tint	0.59	47
Green tint	0.60	68
Reflective	0.42	26
Double Glazed, low-e		
Clear	0.67	76
Gray tint	0.42	37
Bronze tint	0.44	44
Green tint	0.47	64

by allowing outside air to infiltrate into the facility. Energy must be used by the facility's HVAC system to heat or cool any air that infiltrates into the facility. How much energy that is required depends on the temperature difference between the outdoor air and the air inside the facility, and the rate at which air is entering the facility. Of the three ways, infiltration is the one that is most underestimated when considering the energy impact of doors and windows. Also, the rate at which air infiltrates is directly influenced by the condition of the doors and windows.

Even the tightest-closing doors and best-sealed windows will allow air to infiltrate into the facility. It is controlled through the use of close tolerances in the manufacture and installation of the units, and through the use of sealants and weather-stripping.

A Special Warning About Overhead Doors

Overhead doors represent a potentially large source of energy loss for facilities, as a result of both poor maintenance practices and poor operating practices. Overhead doors are subject to physical abuse due to impact from equipment and vehicles. Weather-stripping and door seals are easily damaged. Most are poorly insulated. Many are left open for extended periods of time.

To prevent excessive energy losses from overhead doors, they should be inspected on a regular basis for damage and for proper operation. Doors should be cycled to ensure that they close properly. All door seals and weather-stripping should be inspected for damage. Rollers, pulleys, and cables should be inspected for proper operation and lubricated if necessary. Door surfaces should be inspected for cracking and impact damage.

Equally important, operating procedures must be put in place that dictate how the door is to be used. An open overhead door presents a very large opening in a building exterior. During the heating season, heat losses through open doors can exceed the system's ability to supply heat to the space. Similarly, if the space is air conditioned, heat gains through the open door will most likely exceed the system's ability to cool the space.

HOW TO DEVELOP A DOOR AND WINDOW MAINTENANCE PROGRAM

The goal of the door and window maintenance program is to carry out activities that will ensure that doors and windows reach their maximum service life while enhancing the appearance and operation of the facility, all at a minimum energy cost. To achieve this goal, inspections and maintenance activities will have to be performed on the doors and their frames, the windows and their frames, and their connection to the building envelope. The frequency with which these activities are performed will vary with the types of doors and windows installed, how frequently they are used, the level of abuse they are subjected to, and their exposure to the elements. Units that operate successfully for twenty years or more in one facility may need replacement in less than five years in another facility.

Step 1: Develop the Door Inventory

The first step in developing the maintenance program is the completion of an inventory of the installed exterior doors. Before an inspection program can be put in place and before a scheduled maintenance program can be implemented, the maintenance department must have a detailed listing of what doors are installed where in the facility and what type they are. Most facilities simply have too many exterior doors to rely on the memory of maintenance personnel to see that all doors are inspected and maintained on a regular schedule. Maintenance would be spotty at best. Some doors would receive regular maintenance attention while others would be overlooked for years. Sound door maintenance programs must start from a solid foundation that includes all of the installed doors.

Use Figure 12.4 to develop the door inventory. With multiple doors installed at most entrances, it is critical that a system be developed that can identify a particular door, and relate it to both inventory and maintenance data. For that reason, most door maintenance programs assign a unique inventory number to each door in the facility. All maintenance activities performed on a particular door, including both scheduled and breakdown maintenance, should reference the door inventory number. By using the unique inventory number, it will be possible for you to track maintenance requirements of each door in the facility, identifying ones that may be in need of replacement or upgrading to a heavier duty door.

Start the inventory by gathering the data on the high-use doors first. These are the doors that most likely will require the highest level of maintenance support, and are the ones most likely to malfunction and waste energy. Once their information has been recorded, expand the inventory to include all remaining doors in the facility. It is important that all doors be included in the program as even one improperly operating door can significantly increase energy requirements.

Completing an inventory will also help to identify how much effort will be required to properly maintain the facility's doors. The actual number of doors throughout a facility may not be apparent until the inventory is completed. It is not uncommon to underestimate the number by 50 percent. An accurate count of doors by type will help in establishing maintenance budgets and work schedules for the maintenance program.

Three Door Maintenance Activities

There are three types of activities essential to the success of a door maintenance program: (1) regular inspections; (2) regular testing of door operation; and (3) scheduled maintenance. The purpose of all three activities is to identify problems as they are developing, well before they become high-cost maintenance items or items that result in the waste of energy through the door.

Figure 12.5 lists recommended inspection, testing, and maintenance activities and their frequencies for doors that are used under normal service conditions. Doors that are used under more severe service conditions, such as those in high-traffic facilities

Figure 12.4
Door Inventory

1. Building: _____

2. Location: _____   3. Inventory number: _____

4. Type of door: _____

5. Orientation: ☐ left hand   ☐ left hand reverse   ☐ right hand   ☐ right hand reverse

6. Height: _____   7. Width: _____

8. Lockset: ☐ mortise cylinder   ☐ unit   ☐ cylindrical   ☐ integralock

9. Closer: ☐ none   ☐ crank & piston   ☐ rotary piston   ☐ rack & pinion   ☐ floor

☐ narrow projection   ☐ semi-mortise   ☐ semi-concealed   ☐ concealed

10. Panic exit mechanism: ☐ none   ☐ rim   ☐ mortise   ☐ exposed vertical

☐ concealed   ☐ vertical rod

11. Weather-stripped: ☐ yes   ☐ no

12. Glazing type: _____

13. Glazing size: _____

14. Date installed: _____

15. Comments: _____

_____

_____

_____

or those that are exposed to adverse weather conditions, will require more frequent attention. Doors in normal service should be inspected twice a year. Doors that are subjected to high traffic should be inspected every three months.

The most important element in the door maintenance program is regular inspections. Inspections are required to determine the condition of door components that would otherwise be overlooked. Remember, just because a door is opening and closing does not mean that it is working properly. Inspections will identify problems that contribute to the long-term deterioration of door components and increase energy losses through the door.

Testing of the operation of the door is important to both the long term maintenance of the door and to reducing energy losses. Regular door testing will help to identify problems such as closers that are operating too quickly or too slowly, doors that are out of alignment, and doors that do not properly or fully close.

Finally, scheduled maintenance activities will help to prolong the life of door components. Hinges will wear less and will last longer if they are properly lubricated. Door closers will perform better if they are adjusted regularly. Door thresholds will be more effective in keeping dirt and water out of the facility if they are cleaned on a regular basis.

Step 2: Develop the Window Inventory

As with the door maintenance program, the first step in establishing a maintenance program for windows is the completion of a window inventory. In order to inspect and maintain the windows, maintenance personnel must know how many windows are installed, what type of windows they are, and where they are installed. Although facilities typically have a large number of windows, each must be inspected and maintained on an individual basis if the program is to identify and correct individual maintenance problems that impact both the function and the energy efficiency of the windows. Simply evaluating the average condition of a bank of windows from outside of the facility will miss identifying what corrective action is required.

Use Figure 12.6 to develop the window inventory. To assist in identifying individual windows that may need maintenance action, most window maintenance programs assign a unique inventory number to each window in the facility. A common numbering system makes use of the building room number and then follows a set pattern for numbering the windows in that room, such as numbering from left to right, or clockwise from the room's doorway. All inspection and maintenance activities for a particular window must identify the window's inventory number to allow tracking of the maintenance history for a particular unit.

Three Window Maintenance Activities

A window maintenance program requires that three activities be completed on a regular basis for all windows in the facility: (1) inspection of the window components; (2) testing

Figure 12.5
Door Maintenance Activities

1. Building: _____

2. Door inventory number: _____

3. Date: _____

Maintenance Tasks

Inspect door frame	☐	_____
Inspect door finish	☐	_____
Check door alignment	☐	_____
Inspect hinges for tightness	☐	_____
Lube hinges	☐	_____
Inspect closer	☐	_____
Test closer operation	☐	_____
Adjust closer	☐	_____
Test lock operation	☐	_____
Inspect lock for wear	☐	_____
Lube lock	☐	_____
Inspect threshold	☐	_____
Inspect weather-stripping	☐	_____
Inspect door glazing	☐	_____

Figure 12.6
Window Inventory

1. Building: _____

2. Location: _____   3. Inventory number: _____

4. Type: ☐ fixed ☐ casement ☐ awning ☐ sliding ☐ double hung ☐ jalousie

5. Construction: ☐ aluminum ☐ steel ☐ vinyl ☐ wood

6. Finish: ☐ painted ☐ anodized ☐ aluminum clad ☐ vinyl clad ☐ galvanized

7. Number of glazings: ☐ one ☐ two ☐ three

8. Separate storm sash: ☐ yes ☐ no

9. Glazing color: ☐ clear ☐ gray ☐ bronze ☐ green ☐ reflective

10. Low-e coating: ☐ yes ☐ no   11. Sash: ☐ fixed ☐ operable

12. Type of weather-stripping: ☐ none ☐ metal ☐ vinyl ☐ foam

13. Height: _____   14. Width: _____

15. Lead-based paint: ☐ yes ☐ no ☐ not tested

16. Year installed: _____

17. Comments: _____

_____

_____

_____

of window operation; and (3) performance of maintenance activities. All three are essential to the success of the program, particularly if the energy losses through windows is to be minimized.

Figure 12.7 lists recommended inspection, testing, and maintenance activities for windows that are used under normal service conditions. Windows that are used under normal service conditions should be inspected once per year. Windows that are used under more severe service conditions, such as those that are subjected to severe weather conditions, will require more frequent attention

As with doors, the window maintenance program emphasizes the inspection of window components. The inspections are designed to identify minor problems as they are developing, well before they become major ones that decrease the life expectancy of the window and increase energy losses. The cost of the inspection and the performance of maintenance activities is generally recovered in energy savings within one to two years. The maintenance program can realistically be expected to nearly double the life expectancy of most windows.

When inspecting windows, particular attention must be paid to the condition of the sealant used between the window frame and the opening in the building wall, and between various window components. Sealants serve two primary purposes: they prevent water from penetrating into window components and the building interior, and they limit the rate at which air can infiltrate the facility. Sealants that have separated, cracked, or sagged are not properly protecting the window and will result in long-term water damage to window frame components and to interior finishes. The best time to inspect for sealant damage is during cooler, cloudy days when the joints between window components and between the window frame and the wall are at their largest.

If the inspection of the windows finds a number of locations where the window sealants have failed or are deteriorating rapidly, it is time to plan for replacement of all of the sealants. Even minor deterioration of window sealants can lead to major repair costs and energy losses if they are not promptly repaired.

IMPLEMENTING A DOOR AND WINDOW MAINTENANCE PROGRAM TO REDUCE ENERGY REQUIREMENTS

If maintenance of doors and windows has been carried out on a breakdown basis for some time, there will be significant start-up costs associated with implementing the door and window maintenance program. Inventory data will have to be collected on doors and windows. Condition assessments will have to be completed. Deficiencies will have to be corrected. In some cases, the units will have deteriorated sufficiently to warrant replacement.

If the number of doors and windows is large, and the required repairs significant, the program can be broken down into two phases each for doors and windows. Separate the door inventory into high- and low-use units, and implement the program on the high-use doors first. These are the doors that are most likely in the worst condition and will provide the greatest benefit of a planned maintenance program. Once these doors have been fully upgraded, expand the program to include the remaining doors.

Figure 12.7
Window Maintenance Activities

1. Building: _____

2. Location: _____ 3. Inventory number: _____

4. Date: _____

Maintenance Tasks

Inspect frame paint ☐ _____

Inspect frame for rot/corrosion ☐ _____

Inspect sash paint ☐ _____

Inspect sash for rot/corrosion ☐ _____

Inspect glazings for cracks ☐ _____

Inspect putty/glazing sealant ☐ _____

Inspect sill paint ☐ _____

Inspect sill for rot/corrosion ☐ _____

Inspect frame/wall sealant ☐ _____

Inspect sash/frame weather seal ☐ _____

Test operation ☐ _____

Check fit of sash in frame ☐ _____

Test lock ☐ _____

Inspect handles ☐ _____

The window inventory also should be split into two phases based on the operability of the windows. Include as many operable windows as possible in the first phase. These are the windows that undergo the most wear and therefore are most likely to have defects that will interfere with their operation and increase energy use. If the facility has only operable windows, start with those having southern or western exposure, as these windows will have been exposed to the more extreme temperatures. Expand the program as quickly as possible to include all of the remaining windows.

The Repair/Replace Decision

Throughout the operation of the maintenance program for doors and windows, maintenance personnel must be constantly evaluating the option of replacing the existing windows against continuing to perform maintenance on them. Doors and windows are both expensive items to replace. For windows, the replacement decision is generally all or none due to the impact that new windows have on the appearance of the facility. Unless there are unusual circumstances concerning a particular window that is beyond the point of economic repair, all windows are generally replaced at the same time.

The situation is somewhat different for doors. Depending on the condition of the existing doors, doors can be replaced individually, in pairs, by building exposure, or all at once. Regardless of which approach is used, replacement is a costly, disruptive process.

When considering the replacement option for both windows and doors, maintenance managers must look at a number of factors, including cost and efficiency. To determine the cost of retaining the existing doors and windows, maintenance records will have to be reviewed to identify the types of problems that have been experienced in the past, their frequency, and if the number of problems is increasing or remaining constant. Cost estimates can be compiled from the maintenance records and compared to the cost estimates for a replacement program.

Costs, though, are only one consideration. In many cases, even if the existing windows or doors were restored to like new condition, their energy use would be much greater than that for new, energy efficient units. New units have lower rates of conduction and infiltration. New windows also can control solar heat gain while reducing radiant losses at night. If the doors and windows are in marginal condition, perform an economic analysis on the energy benefits of replacement.

Environmental factors also must be considered. Older steel and wood doors and windows may have been coated with paints or primers that contain lead. Restoration or replacement of these units will require that the maintenance department adhere to lead abatement regulations that dictate how these units may be treated, stripped, handled, and disposed of. Abatement costs are often overlooked when evaluating options, yet can significantly alter the economics.

SUMMARY

A planned maintenance program for a facility's doors and windows is an expensive undertaking. Start-up costs are high and labor-intensive. The payoff comes in the form of improved appearance of the facility, longer operational lives for the doors and windows, and lower energy costs for the facility.

The basic elements of the program include an inventory of the doors and windows installed in the facility, regularly scheduled inspections, performance of corrective maintenance activities, and an evaluation of replacement options. Although the program may appear expensive on paper, the long-term benefits and cost savings readily justify its expense. There are no alternatives if maintenance managers want to gain the most out of their investment in doors and windows. Simple breakdown maintenance cannot match the results of a comprehensive maintenance program.

Chapter 13

Roofs

This chapter presents guidelines and procedures on the maintenance of building roofs to promote energy conservation. The information presented in this chapter may be applied to the types of roofs typically found on commercial, institutional, and industrial facilities. Maintenance activities, testing procedures, and inspection items are identified for a variety of roof types. These activities may be used in establishing a preventive maintenance program for those roofs to enhance their performance and to reduce heat losses and gains through the roof.

FACILITY ROOF ENERGY MAINTENANCE CONSIDERATIONS

Roofs are the single most important component in a building envelope. They bear the brunt of exposure to the elements. Roofs are exposed to greater environmental extremes than any other envelope component. They are subject to attack from ultraviolet light, heat, rain, snow, hail, wind, pollution, wildlife, mechanical equipment, and maintenance mechanics. Even minor damage to a small area can have major consequences for the life, performance, and energy efficiency of the roof.

More than any other component, roofs determine the energy efficiency of the facility's envelope. In typical low-rise construction, the roof accounts for as much as 80 percent of the envelope area of the entire facility. Roof insulation that is saturated with water as the result of roof leaks will gradually increase heat loss and gain through the roof. If a roof is to be energy efficient, its insulation must be kept dry. If the insulation is to be kept dry, the roof must be properly maintained.

Roofs have long been a headache for maintenance managers. Surveys of maintenance managers show that nearly one-third of all new roofs develop significant problems in their first year of use. More than 90 percent of all roofs have experienced major problems by the time they are five years old. Although roofs represent only a small fraction of the total construction cost for a facility, they are a factor in more than 75 percent of all litigation involving new facility construction.

The need for a thorough roof maintenance program is particularly important when one considers the consequence of roof failures. When a roof fails, the operation of the facility is interrupted, often for weeks. Occupants frequently have to be temporarily relocated while repairs are made. Additional repairs have to be made to the building interior and its contents, repairs that average nearly ten times the cost of the repairs to the roof itself.

FORCES ACTING ON THE ROOF

From the moment a roof is installed it is under attack from a number of different forces. Understanding what these forces are and how they impact roof performance will assist you in establishing a maintenance program designed to enhance roof performance. There are three general classifications of forces acting on the roof: (1) heat and radiation forces; (2) chemical forces; and (3) physical forces.

Heat and Radiation Forces

One of the strongest forces acting on the roof is heat. Heat from the sun raises roof surface temperatures to as high as 140 degrees Fahrenheit. At these elevated temperatures, the rate of evaporation of volatile elements from the roofing materials is increased. Chemical breakdown and oxidation of roofing elements are accelerated. Thermal stresses as the result of cycles of expansion and contraction are increased.

One of the effects that heat has on asphalt-based roof materials is the formation of hairline cracks. When asphalt materials are exposed to the temperatures commonly found on roofs, they form a hardened skin. This skin does not have the same thermal expansion characteristics as the underlying material. As the roof goes through repeated cycles of heating and cooling, the skin is unable to expand and contract at the same rate as the underlying material, resulting in a buildup of stresses. Eventually, these stresses relieve themselves in the formation of a series of small, random cracks. These cracks continue to grow with repeated thermal cycles and will eventually extend through the roof's membrane.

The different materials used in roof construction have different rates of thermal expansion that induce stresses where the materials come in contact with each other. These stresses show up as hairline cracks and splits in the joints between the membrane and the flashing, gravel stops, and roof penetrations. With repeated heating and cooling cycles, these hairline cracks and splits grow, allowing water to penetrate the roof.

The impact of thermal stresses can be reduced by reducing the temperature of the roof's surface. ***Rule of Thumb:*** For roofs that require ballast, use a light-colored stone or paver block to reflect at least some sunlight. For unballasted roofs, apply a light-colored coating that is compatible with the roof membrane.

Another major force that produces changes in roofing materials is ultraviolet light from the sun. Ultraviolet light triggers chemical reactions in roofing materials that alter their properties, including a reduction in tensile strength and flexibility. In both asphalt-based and synthetic-based roofing materials, ultraviolet light breaks down complex chemical chains and accelerates oxidation of the material, altering its physical properties. The process is slow, taking years to be noticed. But by the time noticeable changes have taken place, the damage has already been done. Pitch pockets shrink. Membranes and base flashings deteriorate. Sealants fail. Eventually, the loss of flexibility and tensile strength is sufficient to cause the formation of tears and splits.

Rule of Thumb: Unballasted roofs are protected from the effects of ultraviolet light through the use of protective coatings. These coatings should be inspected regularly and reapplied every three to five years. The membranes of ballasted roofs are protected by the ballast material as long as it is evenly distributed over the membrane. Areas where the ballast is thin or missing receive no protection. Ballasted roofs should be inspected on a regular basis for proper coverage by the ballast material.

Chemical Forces

Chemical forces are slow-acting processes that over the course of years work to deteriorate roofing materials. As the chemical forces work on the materials, the properties of these materials are gradually changed and eventually they fail. The most common chemical forces acting on a roof are those of oxidation, volatilization, and reaction to products in building or equipment exhaust fumes.

Oxidation occurs in nearly all roofing materials from the membrane to the roofing deck. Oxidation of asphalt and coal tar roofing materials causes the material to lose flexibility and tensile strength, resulting in increased damage from thermal and structural movement within the roof. Oxidation of synthetic materials leads to polymerization and crosslinking, both of which contribute to a hardening of the materials and a loss of flexibility. Oxidation of the metal roofing decks weakens the deck, resulting in greater movement of the deck components, separation from the overlying roof materials, and loss of strength. Fasteners, used to tie down the roof membrane, lose strength through oxidation and can fail, resulting in separation of the membrane.

The high temperatures at which roofs normally function accelerate this rate of oxidation. Long-term exposure to water, such as in areas where ponding occurs, also accelerates oxidation. Oxidation rates can be reduced through the use of protective and reflective coatings that help to reduce the surface temperature of the roof.

Volatilization is the evaporation of some chemicals from the roof materials. Most roof materials have low rates of volatility at normal room temperatures. However, some materials used in the manufacture of roofing products do have higher volatility rates, particularly at the temperatures commonly found in the roof environment. Plasticizers, hardeners, and other chemical additives will evaporate over time, particularly when they are at elevated temperatures. The result is a loss of strength and flexibility in the material.

Volatilization rates are determined by the types of materials used in the roof construction and the temperature of the roof. While you can do nothing to change the materials used in the roof construction, you can reduce the temperature of the roof's surface by applying reflective coatings or by ensuring that the roof is properly covered with a light-colored ballast.

Another source of chemical attack on roofing materials is the exhaust from processes that take place within the building. For example, animal and vegetable oil vapors from kitchen exhausts attack and dissolve some single-ply roof membranes. Oil and grease from mechanical equipment mounted on the roof attacks roof membranes, resulting in leaks and early failure. Kitchen exhausts must be discharged away from roof surfaces. Areas around mechanical equipment, particularly those prone to leaks, should be protected through the use of pads, pans, or protective coatings that resist oil and grease.

Physical Forces

The most easy to identify damage to roofs comes from physical forces, yet physical damage is rarely noticed or acted upon quickly. In some cases, the damage is slow to develop, so slow that it is accepted as being normal. In other cases, the damage is localized. Unless the roof is thoroughly inspected on a regular schedule, this localized damage will not be detected. There are four general types of physical damage to roofs: (1) foot traffic; (2) ponding; (3) wind; and (4) movement.

Foot Traffic

Roofs are very convenient places for building designers to locate mechanical and electrical equipment, such as cooling systems, exhaust fans, cooling towers, antennas, and satellite dishes. Maintenance personnel must have routine access to this equipment for service. As a result of this regular foot traffic, paths become worn in the roof's upper layer, ballast is displaced, membranes are worn thin, and ultraviolet protective coatings are destroyed. Additional damage occurs when tools and equipment are dropped on the roof, heavy equipment is dragged across roof surfaces, and overspray from cleaners and degreasers comes in contact with roof membranes.

To protect roofs against damage from foot traffic, all facilities must first limit access to roofs to authorized personnel. Areas that are accessed on a regular basis should be protected by pads and walkways. ***Rule of Thumb:*** Roofs should be inspected for damage immediately following completion of any project that involved replacing major equipment located on the roof.

Ponding

The second type of physical damage to roofs is ponding. Ponding occurs when water fails to drain quickly from an area on a roof. Ponding can result from clogged roof drains, settlement of a portion of the building, or poor roof design. Any area where water remains on a roof more than twenty-four hours after a rain has ended is considered to be ponded.

Ponding causes roof problems in two ways. First, ponding keeps roofing materials in extended contact with water. This prolonged contact increases the chances of water entering the layers of the roof through small imperfections. Repeated freeze–thaw cycles accelerate the growth of these imperfections, allowing even more water to enter the roof. Ponding also promotes deterioration of roofing materials and growth of biological agents that can attack roofing materials.

The second way in which ponding attacks the roof is through deflection. Ponded water represents an additional load on the roof membrane. For example, a one-inch-deep pond covering an area of ten by fifteen feet adds a load of 800 pounds to the roof, enough weight to induce a minor sag in the roof. As the sag increases, it allows more water to pond, further increasing the sag. Eventually, the weight of the ponded water and its induced sag is sufficient to cause cracks and tears in the surface of the roof membrane, allowing water to enter the underlying materials. If the ponding occurs near flashing along a parapet wall or an equipment curb, the weight of the water can cause the roof membrane to pull away from the flashing.

Rule of Thumb: To control the damage caused by ponding, roofs should be inspected regularly after rain storms. If ponded areas are discovered, check the roof drains for proper operation first. If the roof drains are not at fault, it may be necessary to install additional roof drains, relocate existing drains, build up sections of roof, or install temporary pumps.

Wind

Wind is a significant force in causing damage to roofs. As the wind blows across the roof's surface, it induces a pressure change that has potential to lift the membrane away from the structure, just as air flowing over the surface of a wing produces lift. If there is insufficient ballast to keep the membrane attached, if there are insufficient fasteners, or if the fasteners are damaged or loose, the membrane will lift, placing stress on the seams of the membrane. While the winds produced during storms are recognized as being strong enough to cause damage, including pealing back large areas of membrane, lower velocity winds over time also can induce significant damage by stressing fasteners and rearranging ballast.

The risk of wind damage can be minimized by regular inspections of the ballast and roof fasteners. If damage is found, additional ballast should be added and defective fasteners replace immediately. Regularly inspect all perimeter flashing to ensure that it is tightly adhered to the roof to prevent wind damage. *Rule of Thumb:* All roofs should be inspected for wind-induced damage immediately following strong storms. Facilities exposed to high winds on a regular basis should have additional ballast or roof pads installed along the perimeter of the roof.

Movement

There are two types of movement that cause damage to the roof: thermal and structural. Thermal movement, as the result of daily cycles of heating and cooling, induces stresses in roofing materials. As the roof continues to go through the heating and cooling cycles, these stresses result in fatigue failures, including the formation of small cracks and splits. Vibrations from roof mounted mechanical equipment can produce similar fatigue failures.

Structural movement, such as that caused by unequal settlement of the building, typically results in the formation of low spots on the roof, or tears along expansion joints. The low spots are subject to ponding and the tears allow easy access of water into the underlying layers of the roof.

Damage from thermal and structural movement must be detected and corrected as quickly as possible. Although thermal damage can be minimized by using a light-colored finish on the roof or by using light-colored stone as ballast, there is little that can be done to prevent or correct for structural movement. ***Rule of Thumb:*** Structural movement is best dealt with at the time of re-roofing.

HOW TO DEVELOP A ROOF MAINTENANCE PROGRAM

Of the components that make up the building envelope, the roof is the most neglected. It is the classic case of being out of sight and out of mind. When problems develop in other building envelope components, they are readily visible to the building owners, maintenance managers, and the building occupants. When problems develop in building roofs, they go unnoticed until water is discovered in the building interior.

Contributing to the problem is the widespread belief among both building owners and maintenance managers that if a roof is not leaking, it is in good shape. Their approach to roof maintenance is to do nothing until someone or something gets wet. Unfortunately, most roof problems develop long before water even enters the building. By the time the leak appears, major damage will most likely have already occurred to the roof, including saturated insulation. As the number of leaks increases with time, pressure from the building occupants to do something typically results in the installation of a new roof, and the cycle starts over. In many cases, this decision to replace is made without fully understanding the condition of the existing roof.

The "do nothing" approach to roof maintenance is directly responsible for the failure of most roofs well before they reach their 15- or 20-year design lives. Although some roof installations suffer from bad design and poor workmanship problems, the number that fail prematurely as a result of neglected maintenance is much greater. The "do nothing" approach fails to recognize that a roof represents a major investment. The result of the neglect is shortened roof life, higher energy costs, increased damage losses to the facility contents, and increased maintenance costs.

Building roofs begin to deteriorate the moment they are installed. Extreme temperature variations, other environmental factors, roof traffic, and the operation of mechanical equipment stress the roof's components. If you are to gain the most from the investment in the facility's roof, you must implement a comprehensive maintenance program. Organizations that have implemented roof maintenance programs typically find that their roofs last twice as long as they did before the program was put in place.

Additionally, they have experienced a lower number of roof leaks, performed fewer emergency repairs, and experienced less damage to interior finishes and equipment.

Roof maintenance programs are successful and cost-effective because they identify minor problems long before they spread and develop into major expenses. By detecting and correcting problems from the outside before water enters the facility, the programs help to keep the roof insulation dry, reducing heat loss and gain.

One of the most important benefits of the programs is that they provide you with solid information needed to evaluate the condition and remaining life of your roof. With that information, you can lay out a multi-year program for roof replacement. Without this information, roof decisions are based more on budget than on need. Roofs that may need replacement often are patched and limp along for years, increasing maintenance and energy costs and the risk of damage. Roofs that merely need repairs may get replaced simply because there was no real understanding of what needed to be done. Either way, more money will be spent fixing the roof than would have been required to maintain it.

Maintenance programs for roofs are not difficult or expensive to implement. Depending on the type of roof installed, annual costs for the program range between two and ten cents per square foot, while roof replacement projects cost between five and fifteen dollars per square foot. The ongoing maintenance costs are a small price to pay for doubling the service life of a roof.

There are three major components to the programs: the completion and ongoing updating of an inventory of the existing roofing construction; the completion of annual or bi-annual inspections; and the conduction of a roof condition test every second or third year.

Step 1: Develop a Roof Inventory

Documentation is a very important part of a roof maintenance program. Accurate records of what is installed, the types and locations of past roofing problems, and what maintenance activities have been performed on the roof are necessary for tracking roof performance and diagnosing roof problems. The documentation file will assist in the budgeting and planning efforts for roof replacement projects.

A documentation file for each roof should be established as soon as a new roof is installed. Included in the file are as-built plans for the roof, current plans showing all roof penetrations, documentation of all repairs made, documentation of all changes made to the roof, and a copy of the roof warranty. The specific information needed to be gathered will vary with the type of roof installed. Information requirements for each major roof type are discussed later in this chapter.

Step 2: Inspect Roof

The single most important element in any roofing maintenance program is regular and detailed inspection. Roof inspection programs are designed to identify defects while they are small, well before they have developed sufficiently to cause extensive and expensive damage to the roof or the underlying structure. By tracking data collected from these inspections, it will be possible to determine when a roof is approaching the point where further maintenance efforts cannot be justified, and a replacement roof is needed.

All roof inspections should start in the interior of the facility. Ceilings, interior surfaces of outside walls, and the underside of the roofing deck should be inspected for signs of water damage from existing leaks. Make certain that the water is coming from the roof rather than from mechanical equipment or piping. While it is difficult to identify the location of the leak just from observing the areas showing water damage, as the water may have traveled a great distance from the leak to the point where it enters the facility, the inspections will help to determine the extent to which the existing roof is leaking. The deck itself should be inspected from the inside for damage, including rusting, rot, deflection, and cracking.

Once the interior of the facility has been surveyed, the inspection should move on to the roof's surface. But the inspection is not limited to the roof's membrane. Many sources of leaks actually originate in components other than membranes, including flashings, expansion joints, drains, and gravel stops. Early identification of defects in these components can help to prevent their becoming points for water entry.

All information gathered during these inspections should be entered into the roof's history file for future reference. Problems that are found to have expanded with each new inspection are early warning signs that the roof may be nearing the end of its useful service life. Photographs of defects can be helpful in tracking how they change over time. If the roof is still under warranty, photographs will help to demonstrate warranty related problems.

The procedures used to complete the roof inspections vary with the type of roof installed and will be discussed later in detail for each roof type.

Step 3: Conduct Roof Tests

The purpose of roof testing is to provide information on the condition of the roof that cannot be provided by visual inspections. While visual inspections can provide good information on the condition of the membrane, the flashing, and other visible roof components, they cannot provide any information on where and to what extent water has penetrated the underlying components of the roof. The presence of water in underlying roof components can be detected only through a number of specialized testing procedures.

Core sampling has long been used to test for the presence of water in roof materials below the membrane. Core sampling is a destructive test in that it involves cutting holes in the roof and removing samples, which are tested for moisture content. To be effective, samples must be taken from both wet and dry areas. Better test results and a better understanding of the areas of the roof that are wet below the membrane require taking more samples.

The major drawbacks of core sampling are that it is destructive and time consuming.

Every location where a core sample is taken must be carefully patched. Each patched area then represents a potential new leak in the roof. It takes time to cut out the samples and install patches. Depending on the condition of the roof, a large number of samples may be required. But even if a large number of samples is taken, the data that they yield will be from a very limited area on the roof. Conditions frequently vary between core samples, making it difficult to judge the overall condition of the roof based solely on the results of core samples. Thus, core sampling works best when combined with one of the other testing techniques.

There are a number of nondestructive tests that can be used in place of or along with core sampling to provide a more accurate understanding of the extent to which moisture has penetrated the roof. These include electrical capacitance, nuclear detection, and thermal imaging. Not all of these techniques work equally well with all roof types. They can, however, provide a very accurate assessment of the condition of a roof's underlying components. In order to use any of the tests, the roof must be dry.

Electrical capacitance testing is based on the ability of a material to store electrical energy. The capacitance meter, when placed on the roof's surface, measures the dielectric properties of the roofing materials. If there is water present in the material, the dielectric properties of the materials change, resulting in a change in the meter reading. For example, dry roofing materials typically have a dielectric constant between 2 and 4. Water has a dielectric constant of 60 to 80. Roofing materials that contain water will show as values of·10 or higher.

In testing a roof using an electrical capacitance meter, the roof is marked off in a grid pattern and readings taken at the grid intersections and recorded on a roof drawing. For most applications, reading are taken on a five-to-ten-foot grid. More readings can be taken in suspect areas to determine the extent of the water penetration.

The most serious drawback of the electrical capacitance test is that the readings can be impacted by metal, including roof decks, nails, fasteners, and foil. The test is also more sensitive to moisture close to the surface of the roof than it is to moisture close to the roof's deck.

Nuclear detection testing uses a small amount of radioactive material to emit high-speed neutrons aimed at the roof. When the neutrons strike a hydrogen atom, some bounce back at a lower speed and are counted by the instrument's detector. Since water contains hydrogen, reflected neutrons indicate the presence of water in the roofing material. Roofing materials also contain hydrogen, so there will be some background reading for even dry materials. When moisture is present in the materials, the count will increase from the normal background reading.

Nuclear detection testing also uses a grid pattern, with readings taken at the grid points and the values recorded on a roof drawing. Readings are generally taken on a five-to-ten-foot grid spacing. Although the meter cannot determine the quantity of moisture present in the materials, it can easily determine the difference between wet and dry insulation. Nuclear detection testing, like electrical capacitance testing, is more sensitive to water found near the surface of the roof.

Thermal imaging detects moisture in the roof's insulation by sensing the temperature differences in the roof's surface. Since the images are taken at night when a temperature difference between the building's interior and the outside air is at least 20 degrees Fahrenheit, areas with wet insulation will show up as being warmer than those areas having dry insulation. Unlike other testing procedures that must take samples based on a grid, thermal imaging examines the entire roof.

Although thermal imaging is very accurate and reliable, several precautions must be taken in order to obtain accurate results. Ideally there should be no wind when the test is performed as wind will conduct heat away from the roof and give false readings. The ballast installed over the roof's surface should be of uniform thickness. Uneven thicknesses and bare areas will also give false readings of heat loss. There should be no roof exhausts or other heat sources discharging air on or near the roof.

Most roof testing programs use a combination of test methods to determine the extent of moisture penetration of a particular roof. For example, any of the nondestructive tests may be used to identify areas where water penetration is suspected, then core sampling may be used to confirm the presence of water and to quantify the amount of water that exists in the insulation.

SIX TYPES OF ROOFS

A number of different types of roofs have evolved over the years to meet the structural and architectural requirements of facilities. In order to properly maintain these different roof type, must be tailored for the particular type of roof installed. To better understand the maintenance requirements of each different roof type, the you must first develop an understanding of how that particular type of roof is constructed.

There are two categories of roofs—flat and sloped. *Flat roofs* are never really flat. Most have a slope of approximately one-quarter inch per foot for drainage purposes. Years ago, some were installed with slopes as low as one-eighth of an inch per foot, but problems with ponding have encouraged designers and manufacturers to increase the slope to a minimum of one-quarter inch per foot. *Sloped roofs* are those with a much higher pitch, typically between three and twelve inches per foot.

The most common types of flat roofs include built-up, single-ply, modified bitumen, and metal. The most common types of sloped roofs include shingle and metal.

BUILT-UP ROOFS AND THEIR MAINTENANCE
REQUIREMENTS

Dating back to the mid-nineteenth century, built-up roofs have been one of the most widely used types of roofs in the United States. What has helped to make them so popular is their use of multiple layers of materials for weatherproofing, providing for backup protection in the event one layer is defective or becomes damaged. Properly applied and maintained, built-up roofing can be expected to provide 15 to 25 years of service, depending on the number of layers of felt used.

Built-up roofs may be applied directly to the structural deck of the roof, or they may be applied to rigid board insulation over the structural deck. Structural decks may be wood, steel, or concrete.

There are three major components to the standard built-up roof: (1) the felts; (2) the bitumen; and (3) the surfacing. The felts provide the necessary tensile reinforcement of the roofing materials to resist the pulling forces generated within the roof. A number of different materials have been used to construct the felts, including cotton rag, asbestos, wood pulp, and fiberglass. These felts are installed in a layered format to provide additional strength and to allow the use of more of the bitumen. Roofs are constructed with two, three, or four layers of felt, with the three-layer roof being the most common.

The bitumen mopped into the layers of felt serves to hold the layers of felt together and is the waterproofing element of the roof. The two primary materials that have been used are asphalt bitumen and coal tar pitch. While coal tar resists water more effectively than asphalt bitumen, it melts at a lower temperature, causing it to become plastic at high roof temperatures. When installed on roofs with a pitch of one-half inch or less, this plasticity helps the material to heal minor defects by resealing itself. On higher pitch roofs, it can cause sufficient flow to weaken the integrity of the roof.

Asphalt bitumen does not become plastic at normal roof temperatures, making it better suited for steeper pitch roofs. In addition, the smoother finish of asphalt bitumen roofs makes it easier for maintenance personnel to identify defects.

The most commonly used surfacing on built-up roofs is mineral aggregate. To protect the bitumen from exposure to the sun and damage from foot traffic, gravel or slag is used to cover the entire roof. The material may be applied loose to the roof or it may be imbedded in a still fluid flood coat.

Common Built-up Roof Problems

In order to establish a maintenance program for built-up roofs, it is necessary to understand the most common roof problems that develop. While some problems, such as inadequate slope, can be traced back to the original construction or modification of the roof, most built-up roofing problems are the result of weathering and abuse. There are nine common problems with built-up roofs; bare areas, cracked and split membranes, blisters, fishmouths, ridging, ponding, punctures, alligatoring, and flashing failure. It is important to note that each of these problems results in allowing water to enter the roof membrane and work its way down through the underlying materials. Eventually the water reaches the roof insulation, destroying its insulating properties.

Bare areas occur when wind, rain, and foot traffic rearrange the roof's ballast, exposing the underlying materials to the heat of the sun, ultraviolet light, and damage from foot traffic. To prevent damage to the surface, ballast should be inspected regularly for coverage and thickness.

Cracks and splits in the membrane are caused by a number of different factors. The roof may have been installed without sufficient expansion joints. There may be excessive movement in building sections. Cracking in the underlying substrate may expand through the roof felts. In severely blistered roofs, the blisters may crack open exposing underlying felts. In most cases, cracks are generally discovered only after water has already entered the building.

Blisters are a major cause of failure for built-up roofing. Blistering occurs in the surface of the top layer of bitumen and between layers of felt. Both types of blisters are caused by trapped air or moisture that expands when the roof is heated, or by gaseous emissions from roofing materials that become trapped in the bitumen. While most blisters are small, typically one inch or less in diameter, they can form as large as a foot or more in diameter. Blisters represent weak spots in the roof that are subject to wear and splitting. Once split, they allow water easy access into the underlying layers of the roof.

Fishmouths are open areas that form along the edges of the upper layer of felt as the result of a failure of the adhesion of the bitumen in these areas. Typically caused by poor workmanship during installation of the roof, or by excessive pulling on the roof felts by thermal or structural movement, fishmouths allow water to enter the underlying roof elements.

Ridges and buckles are long, blister-like deformations of the roofing membrane caused by movement of material under the membrane or the warping upward of the insulation board at the joints. Once raised, the surface of the roof is subject to erosion and physical damage.

Ponding, as discussed previously, occurs when water fails to drain properly from the roof. The additional weight of the water causes deformation of the roofing materials, increasing the area subject to ponding. The standing water caused by ponding promotes deterioration of the bitumen and growth of biological agents that attack roof components.

Punctures in most instances, are the result of physical damage caused by users of the roof, maintenance personnel, and falling branches from trees. Punctures can also be caused by fasteners that have failed and backed out through the membrane. One of the advantages of built-up roofs is their use of multiple layers of waterproofing, providing an additional level of protection against leaks. Punctures, however, still occur and may extend through several or all of the waterproofing layers, allowing water to enter the underlying insulation.

Alligatoring is the development of a random pattern of small, surface cracks that give the roof surface the appearance of the skin of an alligator. When the asphalt bitumen in a roof is exposed to the sun, it oxidizes and contracts. This contraction results in the formation of small crevices on the surface of the roof. As the oxidation process continues, the crevices grow deeper, eventually reaching into the roofing felts, allowing water to enter.

Flashing failure is a major problem for built-up roofs. Typically the result of poor workmanship during installation, flashing failures occur when the flashing pulls away from the vertical surface, fails to fully adhere to the roof surface, or tears at joints. Damage to capping materials on parapet walls can allow water to enter the roof membrane behind the flashing. Nearly all flashing failures allow large quantities of water to enter the roof.

Built-up Roof Inventory

The first step in a maintenance program for built-up roofs is the completion of an inventory of all installed built-up roofs. Before an inspection program can be developed and before a scheduled maintenance program can be implemented, the maintenance department must have a detailed record of where built-up roofs are installed, and what types of materials are used in those roofs. Incompatibilities between different roofing materials requires that the maintenance department know as much as possible about the existing roof before repairs are attempted.

Use Figure 13.1 to develop the built-up roof inventory. Use a separate form for every section of a roof on a facility. In addition to the inventory form, complete a sketch for each roof section showing the outline of the roof and the location of all roof penetrations. When defects are found during a roof inspection, they should be noted on the roof to help in future problem diagnostics.

Completing the inventory will help to assure that each roof section is included in the inspection and maintenance program. Identifying the total roof inventory will assist you in planning your roofing maintenance budget and scheduling the roofs that are in poor condition for replacement.

Built-up Roof Maintenance Activities

There are three activities that must be performed on a regular basis if roofs are to perform effectively over their rated lives: inspection, testing, and repairs. The purpose of the inspection and testing programs are to identify areas in which problems are developing before they expand into larger problems. Repairs should be made as quickly as possible using properly trained in-house maintenance personnel, or a qualified contractor. Delaying or deferring repairs will result in more damage and an increase in chances that water will work its way into the roof insulation.

Inspecting of Built-up Roofs

The most important element in the roof maintenance program is regular inspection. All built-up roofs should be inspected twice a year. One inspection should be scheduled

Figure 13.1
Built-up Roof Inventory

1. Building: _____ 2. Section #: _____

3. Area (sq ft): _____ 4. Year installed: _____

5. Manufacturer: _____ 6. Installer: _____

7. Slope of roof:
   Measure the average overall slope of the roof, excluding areas around roof drains.

   ☐ flat   ☐ pitched: _____ inches per foot

8. Deck:
   From the roof drawings or by site inspection, determine the deck structural material.

   ☐ concrete, poured   ☐ gypsum slab
   ☐ concrete, precast   ☐ plywood
   ☐ gypsum plank   ☐ steel
   ☐ other _____

9. Vapor barrier:
   From the roof drawings, determine the type of material used to form the vapor barrier on top of the roof deck.

   ☐ plastic   ☐ roofing paper
   ☐ none   ☐ other: _____

10. Average insulation thickness: _____

11. Method of attachment:
    From the roof drawings, determine the method used to attach the insulation to the roof deck.

    ☐ fully adhered   ☐ mechanical fasteners
    ☐ none

12. Insulation type:
    From the roof drawings, determine the type of insulation installed.

    ☐ fiberglass   ☐ perlite
    ☐ foam glass board   ☐ polystyrene
    ☐ mineral fiberboard   ☐ other _____

13. Number of plies (typically 2-4): _____

Figure 13.1
Built-up Roof Inventory *(continued)*

14. Felt material:
    From the roof drawings, determine the material used for the roofing felts.

    ☐ asbestos   ☐ fiberglass
    ☐ cotton   ☐ other _____

15. Waterproofing:
    From the roof drawings, determine the waterproofing material used.

    ☐ asphalt   ☐ coal tar
    ☐ other _____

16. Ballast:
    Determine the type of ballast installed on the roof.

    ☐ gravel   ☐ pavers
    ☐ mineral granules ☐ slag
    ☐ other: _____

17. Flashing:
    Determine the materials used for the flashings. Check all that apply.

    ☐ aluminum   ☐ PVC
    ☐ copper   ☐ stainless steel
    ☐ galvanized steel ☐ other _____

18. Type of drainage:
    Determine the type of drainage system used for the roof.

    ☐ built-in   ☐ external

19. Number of roof drains: _____

20. Comments: _____

    _____

    _____

    _____

21. Date completed: _____

Figure 13.2
Built-up Roof Condition

1. Building: _____ 2. Roof section #: _____

3. Date inspected: _____

4. Roof deck underside:
Overall condition	☐	good	☐	fair	☐ poor
Rusting or rot	☐	none	☐	minor	☐ major
Signs of leaks	☐	none	☐	minor	☐ major
Sagging	☐	none	☐	minor	☐ major
Uneven joints	☐	none	☐	minor	☐ major

5. Interior ceilings:
Water stains	☐	none	☐	minor	☐ major

6. Interior walls:
Water stains	☐	none	☐	minor	☐ major
Efflorescence	☐	none	☐	minor	☐ major

7. Ballast:
Coverage	☐	good	☐	fair	☐ poor
Condition	☐	good	☐	fair	☐ poor

8. Membrane surface:
Overall condition	☐	good	☐	fair	☐ poor
Alligatoring	☐	none	☐	minor	☐ major
Blistering	☐	none	☐	minor	☐ major
Buckling	☐	none	☐	minor	☐ major
Cracking	☐	none	☐	minor	☐ major
Fishmouths	☐	none	☐	minor	☐ major
Punctures	☐	none	☐	minor	☐ major
Ponding	☐	none	☐	minor	☐ major

9. Felts:
Overall condition	☐	good	☐	fair	☐ poor
Exposed felts	☐	none	☐	minor	☐ major
Separations	☐	none	☐	minor	☐ major
Curling	☐	none	☐	minor	☐ major

Figure 13.2
Built-up Roof Condition *(continued)*

10. Drains:
Overall condition	☐ good	☐ fair	☐ poor
Free of debris	☐ yes	☐ no	
Proper height	☐ yes	☐ no	
Properly clamped	☐ yes	☐ no	

11. Expansion joints:
Overall condition	☐ good	☐ fair	☐ poor
Open joints	☐ none	☐ minor	☐ major
Secured	☐ none	☐ minor	☐ major
Splits or tears	☐ none	☐ minor	☐ major

12. Flashing:
Overall condition	☐ good	☐ fair	☐ poor
Corrosion	☐ none	☐ minor	☐ major
Detached from wall	☐ none	☐ minor	☐ major
Open joints	☐ none	☐ minor	☐ major
Ridging	☐ none	☐ minor	☐ major

13. Coping:
Overall condition	☐ good	☐ fair	☐ poor
Detached from wall	☐ none	☐ minor	☐ major
Rusted	☐ none	☐ minor	☐ major
Open joints	☐ none	☐ minor	☐ major

14. Parapet walls:
Overall condition	☐ good	☐ fair	☐ poor
Cracks	☐ none	☐ minor	☐ major
Failed joints	☐ none	☐ minor	☐ major
Spalling	☐ none	☐ minor	☐ major
Efflorescence	☐ none	☐ minor	☐ major

15. Roof penetrations:
Overall condition	☐ good	☐ fair	☐ poor
Flashing condition	☐ good	☐ fair	☐ poor
Properly sealed	☐ yes	☐ no	
Open seams	☐ good	☐ fair	☐ poor
Physical damage	☐ good	☐ fair	☐ poor

16. Comments: _____

_____

_____

for early spring to identify any damage suffered during the winter months. The other inspection should be scheduled for mid-fall, after the roof has been exposed to the high temperatures of the summer, yet early enough to allow maintenance personnel to make repairs to areas found to be damaged before winter sets in. Additional inspections should be conducted following severe storms that include potentially damaging winds or hail.

Figure 13.2 lists recommended inspection activities for built-up roofs. A separate inspection form should be completed for each section of roof on a facility. When defects are found during the inspection, they should be noted on the drawing of the roof.

Testing of Built-up Roofs

In addition to the bi-annual roof inspections, all built-up roofs should be tested every two or three years. Testing will help to identify the extent to which water has penetrated the membrane and soaked the insulation. Tests that may be used include core sampling, thermographic scans, electrical capacitance gauges, and nuclear gauges.

It is important that the test results be recorded for comparison with past and future test results, so that trends can be determined. *Rule of Thumb:* If more than 25 percent of the roof insulation is found to be wet, the roof is in need of replacement.

Repairing Built-up Roofs

One of the most important elements in repairing built-up roofs is prompt action. Frequent roof inspections and tests will identify problems when they are still minor. The only way to prevent them from developing into larger and more expensive problems is to repair them as quickly as possible. Small punctures and tears in the surface will allow water to penetrate the roof, causing widespread damage to underlying layers. Even if weahter conditions, such as extreme cold, do not allow for permanent repairs to be made, it is critical that temporary repairs be made immediately with the understanding that more permanent repairs be made when weather conditions permit.

Equally important with prompt action is the need to use the proper materials. Incompatibilities between materials used to repair the roof and materials in the existing roof prevent proper bonding of the materials and sealing of the roof. In some cases, incompatibilities will result in chemical reactions that will further damage the roof. Before any repairs are made, check with the roof manufacturer to determine the proper types of materials to be used on that particular roof.

SINGLE-PLY ROOFS AND THEIR MAINTENANCE REQUIREMENTS

Single-ply roofs were introduced during the late 1950s and early 1960s, but were not widely accepted until the 1980s when they were promoted as a low-cost alternative to petroleum-based built-up roofs. Today they have gained acceptance as a result of their ease of installation and their ability to accommodate building movement. This ability to stretch and match building thermal movement has made the single-ply roof particularly well suited for use in areas that experience large swings in temperature. As a result, approximately one-half of all new and replacement roofs that are being installed today on low-sloped buildings are single-ply.

One of the biggest factors that has led to the acceptance of single-ply roofs is the speed with which they can be installed. While built-up roofing must be fabricated in the field, single-ply membranes are manufactured and delivered to the site ready for installation in a variety of widths, greatly speeding installation. In cases where the building structure permits, a single-ply membrane may be applied directly over an existing built-up roof.

The biggest drawback of single-ply roofs is their lack of redundancy. Unlike built-up roofs that have three or four waterproofing layers, singly ply roofing has only the single membrane. Built-up roofs can suffer damage to one or more of the plys without resulting in a leak. When the membrane is damaged in a single-ply roof application, water will gain access to the underlying insulation.

In single-ply roof construction, the membrane is between 30 and 60 mils thick. Depending on the type of materials used to make the membrane, seams may be solvent welded, hot-air welded, or joined with adhesive. Membranes may be applied loose laid with ballast, partially attached, mechanically fastened, or fully adhered. Loose laid membranes offer the advantage of being totally isolated from structural movement and settlement of the building. Loose laid membranes are secured with stone or roofing pavers, typically at ten pounds of ballast per square foot of roof. The ballast must be well-rounded stone, having no sharp corners that could puncture the membrane. In maintaining loose laid membranes, it is essential that the roof membrane be fully covered. If the ballast becomes displaced, the roof membrane can be displaced under high wind conditions. The ballast also serves as a protective layer, shielding the membrane from the sun's ultraviolet light.

Partially adhered membranes are attached with a series of strips or plate fasteners to the roof's supporting structure. Since the membrane is adhered to the substrate, no ballast is required. It is important, however, that the fasteners be inspected on a regular basis. Failed fasteners allow the roof membrane to balloon under high wind conditions, leading to tearing. It is equally important that the membrane be protected from ultraviolet light by an approved coating. Areas subject to foot traffic must be protected with pavers or walkways to prevent wear on the membrane.

Fully adhered membranes are continuously attached to the roof's insulation by contact cement, cold adhesives, or hot asphalt. These membranes must be periodically inspected, ideally during windy conditions, to ensure that the membrane is still adhered to the insulation. Failed adhesion will cause ballooning during windy conditions. Failed adhesion along the perimeter of the roof will cause the membrane to peel away from the roof. Fully adhered membranes must also be protected from ultraviolet light and foot traffic.

There are two classes of materials used in making single-ply membranes—thermoset and thermoplastic. **Thermoset** single-ply membranes are installed in sheets with widths of 20 to 30 feet, making them well suited for roofs having few penetrations. They may be installed fully adhered, partially adhered, or ballasted. Thermoset membranes offer good protection against ultraviolet light and weathering. They are readily damaged by petroleum products, so protective pads must be installed around roof-mounted mechanical systems.

Thermoplastic membranes come in narrower rolls than thermoset, making them better suited for roofs that have a large number of penetrations. Most applications are partially adhered or ballasted. Individual sheets are joined by heat welding. Thermoplastic membranes do not offer good protection against ultraviolet light or petroleum products.

One of the characteristics of thermoplastic membranes is that they contain polymer plastics that are not cross-linked during the manufacturing process. As a result, the membrane can be repeatedly softened using heat in order to make repairs or modifications.

Common Single-Ply Roof Problems

Single-ply roofs, like built-up roofs, suffer a number of problems that if uncorrected will permit water to enter the underlying components of the roof, including the insulation. There are six common problems associated with single-ply roofs, including: (1) seam failure; (2) ponding; (3) fastener failure; (4) alligatoring; (5) membrane punctures; and (6) flashing failures.

Seam failure is the most common problem with single-ply roofs. The causes of seam failures include improper installation techniques, failure of the heat weld, and excessive movement of the substrate. With no underlying protection, failed seams allow water to enter the insulation and the building. The best protection against failed seams is using a qualified contractor to install the roof, and regular inspections to identify seams that have failed. Any failed seam must be patched as quickly as possible.

Ponding, although a problem for single-ply roofs, is not as serious a problem as it is for built-up roofs. The flexibility of single-ply membranes provides better resistance to the effects of ponding. Most instances of ponding can be corrected only at the time of roof replacement. If areas of ponding are found on the roof, check to see that the roof drains are free of debris. Although single-ply membrane materials resist water, ponding can create conditions where the membrane is subject to biological attack, resulting in crazing of the surface.

Fastener failure is a serious problem for single-ply roofs. Insufficient ballast, corroded fasteners, fasteners that have lost their holding power, and fasteners that have backed out of the substrate all result in a membrane that is not sufficiently adhered to the roof. High winds can then peal the membrane back from the corners of the roof, or they can cause the membrane to balloon, stressing and tearing seams. Roofs must be inspected regularly for proper adhesion to the roof substrate, and additional ballast or fasteners installed as required.

Alligatoring. Patterns in the surface may be formed if the roof membrane is not properly protected from ultraviolet light. The membrane will dry out and crack, forming an alligator pattern in the surface. If not corrected, the cracking will continue through the membrane, eventually resulting in its failure. To protect against alligatoring, inspect ballasted single-ply roofs for full coverage. Non-ballasted roofs should have a protective coating applied every three to five years, depending on exposure.

Punctures allow water to quickly gain access to the underlying insulation. With only one layer of protection in single-ply roofs, it is critical that the membrane remain in tact. Most punctures come as the result of foot traffic on the roof. To protect against punctures, any area that is accessed more than once a month should be protected by installing a walkway. All areas surrounding equipment installed on the roof should be inspected closely for punctures and other membrane damage on a regular basis.

All flashings and counterflashings must be carefully inspected on single-ply roofs. Although the membrane of single-ply roofs is more flexible than the membrane installed on built-up roofs, it can concentrate stresses along flashings, resulting in tears and gaps. Damage to capping materials on parapet walls can allow water to enter the roof membrane behind the flashing.

Single-Ply Roof Inventory

The first step in a maintenance program for single-ply roofs is the completion of an inventory of all installed single-ply roofs. The maintenance department must have a detailed record of where the roofs are installed, what types of single-ply roofs are installed, and what types of membranes are used in those roofs. Incompatibilities between materials used in different single-ply roofs require that the maintenance department know as much as possible about the existing roof before repairs are attempted. For example, the materials used to make repairs on thermoset membranes are different from those used to make repairs on thermoplastic membranes. Using the wrong repair materials will not provide adequate adhesion and may damage the membrane.

Use Figure 13.3 to develop the single-ply roof inventory. Use a separate form for each section of a roof on a facility. In addition to the inventory form, a sketch should be completed for each roof section showing the outline of the roof and the location of all roof penetrations. When defects are found during a roof inspection, they should be noted on the roof to help in future problem diagnostics.

Completing the inventory will lay the foundation for the roof inspection program and will help to ensure that each roof section is included. Identifying the total roof inventory will assist you in planning your roofing maintenance budget and scheduling the roofs that are in poor condition for replacement.

Figure 13.3
Single-Ply Roof Inventory

1. Building: _____ 2. Roof section #: _____

3. Area (sq ft): _____ 4. Year installed: _____

5. Manufacturer: _____ 6. Installer: _____

7. Slope of roof:
   Measure the average overall slope of the roof, excluding areas around roof drains.

   ☐ flat   ☐ pitched: _____ inches per foot

8. Deck:
   From the roof drawings or by site inspection, determine the deck structural material.

   ☐ concrete, poured   ☐ gypsum slab
   ☐ concrete, slab   ☐ plywood
   ☐ gypsum plank   ☐ steel
   ☐ other _____

9. Vapor barrier:
   From the roof drawings, determine the type of material used to form the vapor barrier on top of the roof deck.

   ☐ plastic   ☐ roofing paper
   ☐ none   ☐ other: _____

10. Average insulation thickness: _____

11. Method of attachment: _____
    From the roof drawings determine the method used to attach the insulation to the roof deck.

    ☐ fully adhered   ☐ mechanical fasteners
    ☐ none

12. Insulation type:
    From the roof drawings, determine the type of insulation installed.

    ☐ fiberglass   ☐ perlite
    ☐ foam glass board   ☐ polystyrene
    ☐ mineral fiberboard ☐ other _____

Figure 13.3
Single-Ply Roof Inventory *(continued)*

13. Membrane type:
    From the roof drawings, determine the type of membrane installed on the roof.

    ☐ CPE   ☐ Hypalon
    ☐ EPDM   ☐ Neoprene
    ☐ PIB   ☐ PVC

14. Membrane attachment method:
    From the roof drawings, determine the method used to attach the membrane.

    ☐ adhesives   ☐ pavers
    ☐ ballast   ☐ plate fasteners
    ☐ other _____

15. Surface coating: ☐ no   ☐ yes, type: _____

15. Flashing:
    Determine the materials used for the flashings. Check all that apply.

    ☐ aluminum   ☐ PVC
    ☐ copper   ☐ stainless steel
    ☐ galvanized steel   ☐ other _____

16. Type of drainage:
    Determine the type of drainage system used for the roof.

    ☐ built-in   ☐ external

17. Number of roof drains: _____

18. Comments: _____

    _____

    _____

    _____

19. Date completed: _____

Single-Ply Roof Maintenance Activities

There are three activities that must be performed on a regular basis: inspection, testing, and repairs. The inspection and testing programs are very similar to those established for built-up roofs. They too are designed to identify areas where problems are developing before they expand into larger problems. Quick recognition is particularly important for single-ply roofs due to the lack of multiple layers of waterproofing. All repairs should be made as quickly as possible using the materials and procedures recommended by the roofing membrane manufacturer. Delaying or deferring repairs will result in more damage and an increase in chances that water will work its way into the roof insulation.

Inspecting of Single-Ply Roofs

The most important element in the roof maintenance program for single-ply roofs is regular inspection. All single-ply roofs should be inspected at least twice a year. One inspection should be scheduled for early spring to identify any damage suffered during the winter months. The second inspection should be scheduled for mid-fall, after the roof has been exposed to the high temperatures of the summer, yet early enough to allow making repairs to areas found to be damaged before winter sets in. If it is suspected that there are adhesion problems for a particular roof, it may be necessary to inspect how well the membrane is adhered to the substrate during a day when there are relatively high winds. If the adhesion or fasteners have failed, the membrane will balloon during high winds. Additional inspections should be conducted following severe storms that include potentially damaging winds or hail.

Figure 13.4 lists recommended inspection activities for single-ply roofs. A separate inspection form should be completed for each section of roof on a facility. When defects are found during the inspection, they should be noted on the drawing of the roof.

Testing of Single-Ply Roofs

In addition to the bi-annual roof inspections, all single-ply roofs should be tested every two or three years. Testing will help to identify the extent to which water has penetrated the membrane, and soaked the insulation. Tests that may be used include core sampling, thermographic scans, electrical capacitance gauges, and nuclear gauges. Thermographic scans are more accurate on unballasted roofs than they are on roofs that have ballast.

It is important that the test results be recorded for comparison with past and future test results, so that trends can be determined. ***Rule of Thumb:*** If more than 25 percent of the roof insulation is found to be wet, the roof is in need of replacement.

MODIFIED BITUMEN ROOFS AND THEIR MAINTENANCE REQUIREMENTS

Modified bitumen roofs are a cross between built-up and single-ply roofs. They offer the advantage of the multiple plies found in built-up roofing, and the installation ease

Figure 13.4
Single-Ply Roof Condition

1. Building: _____ 2. Roof section #: _____

3. Date inspected: _____

4. Roof deck underside:
   - Overall condition   ☐ good   ☐ fair   ☐ poor
   - Rusting or rot   ☐ none   ☐ minor   ☐ major
   - Signs of leaks   ☐ none   ☐ minor   ☐ major
   - Sagging   ☐ none   ☐ minor   ☐ major
   - Uneven joints   ☐ none   ☐ minor   ☐ major

5. Interior ceilings:
   - Water stains   ☐ none   ☐ minor   ☐ major

6. Interior walls:
   - Water stains   ☐ none   ☐ minor   ☐ major
   - Efflorescence   ☐ none   ☐ minor   ☐ major

7. Insulation:
   - Shrinkage   ☐ none   ☐ minor   ☐ major
   - Expansion   ☐ none   ☐ minor   ☐ major
   - Crushing   ☐ none   ☐ minor   ☐ major
   - Warpage   ☐ none   ☐ minor   ☐ major

8. Membrane:
   - Overall condition   ☐ good   ☐ fair   ☐ poor
   - Alligatoring   ☐ none   ☐ minor   ☐ major
   - Blistering   ☐ none   ☐ minor   ☐ major
   - Buckling   ☐ none   ☐ minor   ☐ major
   - Cracking   ☐ none   ☐ minor   ☐ major
   - Fishmouths   ☐ none   ☐ minor   ☐ major
   - Punctures   ☐ none   ☐ minor   ☐ major
   - Ponding   ☐ none   ☐ minor   ☐ major
   - Ballooning   ☐ none   ☐ minor   ☐ major

9. Fasteners: ☐ n/a
   - Corrosion   ☐ none   ☐ minor   ☐ major
   - Back-out   ☐ none   ☐ minor   ☐ major
   - Loss of holding   ☐ none   ☐ minor   ☐ major

Figure 13.4
Single-Ply Roof Condition *(continued)*

10. Adhesive:
    - Failure at seams   ☐ none   ☐ minor   ☐ major
    - Open area failure   ☐ none   ☐ minor   ☐ major

11. Ballast/pavers:
    - Coverage   ☐ good   ☐ fair   ☐ poor
    - Condition   ☐ good   ☐ fair   ☐ poor

12. Drains:
    - Overall condition   ☐ good   ☐ fair   ☐ poor
    - Proper height   ☐ yes   ☐ no
    - Properly clamped   ☐ yes   ☐ no
    - Free of debris   ☐ yes   ☐ no

13. Expansion joints:
    - Overall condition   ☐ good   ☐ fair   ☐ poor
    - Open joints   ☐ none   ☐ minor   ☐ major
    - Secured   ☐ none   ☐ minor   ☐ major
    - Splits or tears   ☐ none   ☐ minor   ☐ major

14. Flashing:
    - Overall condition   ☐ good   ☐ fair   ☐ poor
    - Corrosion   ☐ none   ☐ minor   ☐ major
    - Detached from wall   ☐ none   ☐ minor   ☐ major
    - Open joints   ☐ none   ☐ minor   ☐ major
    - Ridging   ☐ none   ☐ minor   ☐ major

15. Coping:
    - Overall condition   ☐ good   ☐ fair   ☐ poor
    - Detached from wall   ☐ none   ☐ minor   ☐ major
    - Rusted   ☐ none   ☐ minor   ☐ major
    - Open joints   ☐ none   ☐ minor   ☐ major

16. Parapet walls:
    - Overall condition   ☐ good   ☐ fair   ☐ poor
    - Cracks   ☐ none   ☐ minor   ☐ major
    - Failed joints   ☐ none   ☐ minor   ☐ major
    - Spalling   ☐ none   ☐ minor   ☐ major
    - Efflorescence   ☐ none   ☐ minor   ☐ major

Figure 13.4
Single-Ply Roof Condition *(continued)*

17. Roof penetrations:

Overall condition	☐ good	☐ fair	☐ poor
Flashing condition	☐ good	☐ fair	☐ poor
Properly sealed	☐ yes	☐ no	
Open seams	☐ good	☐ fair	☐ poor
Physical damage	☐ good	☐ fair	☐ poor

18. Comments: _____

_____

_____

_____

of single-ply. Introduced in the United States during the early 1970s, modified bitumen roofs have gradually gained in acceptance and market share.

A modified bitumen roof has three elements: (1) the structural roof deck; (2) the insulation; and (3) a composite sheet membrane. The membrane is a multi-layer sheet consisting of asphalt with polyester or fiberglass reinforcement. Laminated at the manufacturing plant, the membrane is applied like a single-ply roof membrane. Seams are bonded with adhesives or by heat torching. The membrane can be fully adhered or ballasted. Modified bitumen roof membranes are well suited for use in applications that are subjected to large volumes of foot traffic and need puncture resistance.

Common Modified Bitumen Roof Problems

Problems experienced with modified bitumen roofs are similar to those experienced with single-ply roofs. Many are the result of improper installation practices, made worse by weathering. There are four common problems with modified bitumen roofs: (1) seam failures; (2) ponding; (3) inadequate attachment or ballasting; and (4) failed flashing.

Failure of the seams in a modified bitumen roof are generally the result of improper installation techniques including overstressing the seam, inadequate heat, or inadequate pressure. Failed seams can be repaired by applying a fiberglass patch and reinforcing cement to the area. Failed seams must be repaired as quickly as possible to prevent the entry of water into the underlying insulation.

Ponding, resulting from inadequate slope or improper roof drainage, places additional stress on the roof's seams and will eventually lead to failure of the seam. Roofs must be regularly inspected for ponding. While little can be done to correct ponding except at the time of roof replacement, maintenance personnel may find it necessary to use portable pumps to remove ponded water from the roof, thus limiting the damage. If ponding occurs regularly in several areas, the number of inspections completed for the roof should be increased, paying particular attention to the seams in the area of the ponding.

Insufficient ballast will allow the roof membrane to balloon under high wind conditions. In ballasted roof construction, ballast must be evenly spread at the recommended level, typically ten pounds per square foot. In mechanically fastened roofs, failure of the fasteners or an insufficient number of fasteners will also allow the roof membrane to balloon. Regular inspection of the ballast or fasteners is the best defense against wind damage to the membrane. For buildings that are regularly subjected to high winds, additional ballast may be required along the roof's perimeter to lessen the chances of damage.

Failed flashing is a common problem, as with other roof types. Failed flashing often allows water to enter the roof directly below the membrane. Since there is only one membrane in a modified bitumen roof, any flashing failure will allow water to enter the roof's insulation. Particular attention must be paid to flashing during the regular roof inspections.

Modified Bitumen Roof Inventory

The maintenance program for modified bitumen roofs begins with the completion of an inventory of all installed modified bitumen roofs. The maintenance department must have a detailed record of where the roofs are installed, what types of modified bitumen roof are installed, and what types of materials are used in the membranes. Incompatibilities between different membrane materials requires that the maintenance department know as much as possible about the existing roof before repairs are attempted.

Use Figure 13.5 to develop the modified bitumen roof inventory. A separate form should be used for every section of roof on a facility. In addition to the inventory form, a sketch should be completed for each roof section showing the outline of the roof and the location of all roof penetrations. When defects are found during a roof inspection, they should be noted on the roof to help in future problem diagnostics.

Modified Bitumen Roof Maintenance Activities

Modified bitumen roof maintenance activities are similar to those required for single-ply roofs. As with other roof types, there are three maintenance activities that must be performed on a regular basis: inspection, testing, and repairs. With only a single membrane to keep water out, all repairs should be completed as quickly as possible to prevent extending the damage to the underlying roofing materials, including the insulation.

Inspecting of Modified Bitumen Roofs

Regular inspection of the modified bitumen roof is an important factor in maintaining the roof. All roofs should be inspected twice a year, once in the spring and again in the fall. If ponding is an ongoing problem, additional spot inspections of all seams in the area of the ponding should be completed on a monthly basis. Additional inspections should be conducted following severe storms that include potentially damaging winds or hail.

Figure 13.6 lists recommended inspection activities for modified bitumen roofs. A separate inspection form should be completed for each section of roof on a facility. When defects are found during the inspection, they should be noted on the drawing of the roof.

Testing of Modified Bitumen Roofs

The same test methods used on single-ply roofs may be used on modified bitumen roofs: core sampling, thermographic scans, electrical capacitance gauges, and nuclear gauges. Thermographic scans are more accurate on unballasted roofs than they are on

Figure 13.5
Modified Bitumen Roof Inventory

1. Building: _____ 2. Section #: _____

3. Area (sq ft): _____ 4. Year installed: _____

5. Manufacturer: _____ 6. Installer: _____

7. Slope of roof:
   Measure the average overall slope of the roof, excluding areas around roof drains.

   ☐ flat   ☐ pitched: _____ inches per foot

8. Deck:
   From the roof drawings or by site inspection, determine the deck structural material.

   ☐ concrete, poured   ☐ gypsum slab
   ☐ concrete, slab   ☐ plywood
   ☐ gypsum plank   ☐ steel
   ☐ other _____

9. Vapor barrier:
   From the roof drawings, determine the type of material used to form the vapor barrier on top of the roof deck.

   ☐ plastic   ☐ roofing paper
   ☐ none   ☐ other: _____

10. Average insulation thickness: _____

11. Method of attachment:
    From the roof drawings, determine the method used to attach the insulation to the roof deck.

    ☐ fully adhered   ☐ mechanical fasteners
    ☐ none

12. Insulation type:
    From the roof drawings, determine the type of insulation installed.

    ☐ fiberglass   ☐ perlite
    ☐ foam glass board   ☐ polystyrene
    ☐ mineral fiberboard ☐ other _____

Figure 13.5
Modified Bitumen Roof Inventory *(continued)*

13. Membrane:
    From the roof drawings, determine the type of membrane installed on the roof.

    ☐ polypropylene   ☐ rubberized

14. Attachment:
    From the roof drawings or by inspection, determine the method used to attach the membrane.

    ☐ ballasted   ☐ mechanically fastened
    ☐ fully adhered   ☐ pavers

15. Flashing:
    Determine the type of materials used for the flashings. Check all that apply.

    ☐ aluminum   ☐ PVC
    ☐ copper   ☐ stainless steel
    ☐ galvanized steel   ☐ other _____

16. Type of drainage:
    Determine the type of drainage system used for the roof.

    ☐ built-in   ☐ external

17. Number of roof drains: _____

18. Comments: _____

    _____

    _____

    _____

19. Date completed: _____

Figure 13.6
Modified Bitumen Roof Condition

1. Building: _____ 2. Roof section #: _____

3. Date inspected: _____

4. Roof deck underside:

Overall condition	☐ good	☐ fair	☐ poor
Rusting or rot	☐ none	☐ minor	☐ major
Signs of leaks	☐ none	☐ minor	☐ major
Sagging	☐ none	☐ minor	☐ major
Uneven joints	☐ none	☐ minor	☐ major

5. Interior ceilings:

Water stains	☐ none	☐ minor	☐ major

6. Interior walls:

Water stains	☐ none	☐ minor	☐ major
Efflorescence	☐ none	☐ minor	☐ major

7. Insulation:

Shrinkage	☐ none	☐ minor	☐ major
Expansion	☐ none	☐ minor	☐ major
Crushing	☐ none	☐ minor	☐ major
Warpage	☐ none	☐ minor	☐ major

8. Membrane:

Overall condition	☐ good	☐ fair	☐ poor
Failed seams	☐ none	☐ minor	☐ major
Alligatoring	☐ none	☐ minor	☐ major
Blistering	☐ none	☐ minor	☐ major
Buckling	☐ none	☐ minor	☐ major
Cracking	☐ none	☐ minor	☐ major
Fishmouths	☐ none	☐ minor	☐ major
Punctures	☐ none	☐ minor	☐ major
Ponding	☐ none	☐ minor	☐ major
Ballooning	☐ none	☐ minor	☐ major

9. Fasteners: ☐ n/a

Corrosion	☐ none	☐ minor	☐ major
Back-out	☐ none	☐ minor	☐ major
Loss of holding	☐ none	☐ minor	☐ major

Figure 13.6
Modified Bitumen Roof Condition *(continued)*

10. Ballast/pavers:
Coverage	☐ good	☐ fair	☐ poor
Condition	☐ good	☐ fair	☐ poor

11. Drains:
Overall condition	☐ good	☐ fair	☐ poor
Proper height	☐ yes	☐ no	
Properly clamped	☐ yes	☐ no	
Free of debris	☐ yes	☐ no	

12. Expansion joints:
Overall condition	☐ good	☐ fair	☐ poor
Open joints	☐ none	☐ minor	☐ major
Secured	☐ none	☐ minor	☐ major
Splits or tears	☐ none	☐ minor	☐ major

13. Flashing:
Overall condition	☐ good	☐ fair	☐ poor
Corrosion	☐ none	☐ minor	☐ major
Detached from wall	☐ none	☐ minor	☐ major
Open joints	☐ none	☐ minor	☐ major
Ridging	☐ none	☐ minor	☐ major

14. Coping:
Overall condition	☐ good	☐ fair	☐ poor
Detached from wall	☐ none	☐ minor	☐ major
Rusted	☐ none	☐ minor	☐ major
Open joints	☐ none	☐ minor	☐ major

15. Parapet walls:
Overall condition	☐ good	☐ fair	☐ poor
Cracks	☐ none	☐ minor	☐ major
Failed joints	☐ none	☐ minor	☐ major
Spalling	☐ none	☐ minor	☐ major
Efflorescence	☐ none	☐ minor	☐ major

16. Roof penetrations:
Overall condition	☐ good	☐ fair	☐ poor
Flashing condition	☐ good	☐ fair	☐ poor
Properly sealed	☐ yes	☐ no	
Open seams	☐ good	☐ fair	☐ poor
Physical damage	☐ good	☐ fair	☐ poor

Figure 13.6
Modified Bitumen Roof Condition *(continued)*

17. Comments: _____

_____

_____

_____

roofs that have ballast. Testing should be performed every two or three years, unless there are indications that water has penetrated the roof membrane.

All test results should be recorded so that the performance of the roof can be tracked over time. As with other roof types, if more than 25 percent of the roof insulation is found to be wet, the roof is in need of replacement.

METAL ROOFS AND THEIR MAINTENANCE REQUIREMENTS

Initially, metal roofs were used exclusively with metal buildings. However, starting in the 1970s, metal roofs began appearing on other structures. Today, nearly one-half of all new, nonresidential, low-rise buildings use metal roofs. A wide variety of materials and finishes and favorable life-cycle costs have helped to contribute to the increasing use of metal roofs.

There are two classifications of metal roofs—architectural and structural. In architectural metal roof construction, the roof's waterproofing layer and its supporting deck are separate components. The metal waterproofing layer serves as a decorative surface treatment and is not a structurally active element in the roof. The metal roof must be installed over a structurally supported wood or steel substrate, insulation board, and a moisture resistant roofing felt. Architectural metal roofs may be fabricated from steel, zinc copper, stainless steel, or aluminum. The recommended minimum slope is three inches per foot.

In structural metal roof construction, the waterproofing layer and the supporting deck are a single component, the metal portion of the roof. The metal functions as the protective layer against weather and the load distributor for the roof. Metal panels are manufactured from coiled carbon sheet steel and are galvanized, aluminized, or hot dipped aluminum-zinc coated. Panels may be finished with a polyester paint or laminated with a bonded acrylic.

A number of different types of seams are used on metal roofs to join the panels, including flat, batten, and standing. Batten and standing seams offer the advantage of raising the seam several inches above the roof's surface, thus reducing the chances for leaks at the seams. An additional advantage of standing seam roof construction is that the metal panels are free floating. By being independent of the supporting structure, the panels are less likely to suffer thermal stresses.

Common Metal Roof Problems

Due to the different materials and construction techniques used in installing metal roofs, maintenance requirements for metal roofs differ from those for other roof designs. Therefore, in order to develop a maintenance program for metal roofs, it is important to understand common problems that metal roofs experience. There are three common problems experienced with metal roofs: (1) failed or loose fasteners; (2) damaged panels; and (3) corroded panels.

Failed or loose fasteners are by far the most common problem experienced with metal roofs. Fasteners can be damaged in a number of different ways. If they are overtightened during installation, they can be partially stripped. Wind across the roof will induce vibrations that can result in the formation of cracks at the fasteners as the result of metal fatigue. Contact between the fasteners and other roof metals of different types can induce galvanic corrosion. Trapped water in contact with the fastener will promote corrosion. As the fasteners weaken or fail, they will allow the metal roof panels to loosen and become more susceptible to damage from wind.

Damage to metal roof panels may result from a number of things, including high winds, tree branches, and hail. Even maintenance crews performing roof inspections can cause damage. Typical damage includes creased or bent panels, warped panels, tears, dents, and bad seams. If the damage is severe, the panels may pull away from the fasteners, increasing the risk that they will be damaged by high winds.

Corrosion of metal roof panels, including those made of aluminum may occur. The protective coatings applied to the panels are designed to reduce the rate of corrosion, but corrosion will still take place. Regular inspection of the panel finish will identify areas where the finish is no longer properly protecting the metal panels. Periodic recoating of the panels will help to extend their life.

Metal Roof Inventory

The inventory of metal roofs is particularly important considering the many different types of metal roofs that have been installed. Differences in roof construction techniques will result in different maintenance requirements. It is essential to identify as much information as possible on the existing roof construction in order to aid inspection and maintenance efforts.

Use Figure 13.7 to record inventory information for metal roofs. A separate form should be completed for each section of roof on a facility. In addition to the inventory form, a sketch should be completed for each roof section showing the outline of the roof and the location of all roof penetrations. When defects are found during a roof inspection, they should be noted on the roof drawing to help in future problem diagnostics.

Metal Roof Maintenance Activities

The primary maintenance task for metal roofs is the completion of an inspection every spring and fall. During each inspection, particular attention should be paid to the fasteners to ensure they are in good condition and are properly securing the metal panels. Use Figure 13.8 to record inspection information for metal roofs. A separate inspection form should be completed for each section of metal roof on a facility.

Figure 13.7
Metal Roof Inventory

1. Building: _____ 2. Section #: _____

3. Area (sq ft): _____ 4. Year installed: _____

5. Manufacturer: _____ 6. Installer: _____

7. Slope of roof:
   Measure the average overall slope of the roof.

   ☐ flat   ☐ pitched: _____ inches per foot

8. Type of roof: ☐ architectural   ☐ structural

9. Type of seam: ☐ batten   ☐ standing
                   ☐ flat   ☐ other: _____

10. Deck:
    For structural metal roofs, determine the deck structural material.

    ☐ concrete   ☐ metal
    ☐ gypsum plank   ☐ wood
    ☐ gypsum slab   ☐ other _____

11. Roofing material:
    From the roof drawings or by site inspection, determine the material used to construct the roof.

    ☐ aluminum   ☐ steel
    ☐ copper stainless steel   ☐ zinc copper
    ☐ stainless steel

12. Finish:
    From the roof drawings or by site inspection, determine the roof finish.

    ☐ galvanized   ☐ painted
    ☐ natural   ☐ other _____

Figure 13.7
Metal Roof Inventory *(continued)*

13. Flashing:
    By site inspection, determine the type of materials used for the flashings. Check all that apply.

    ☐ aluminum   ☐ PVC
    ☐ copper   ☐ stainless steel
    ☐ galvanized steel   ☐ other _____

14. Comments: _____

    _____

    _____

    _____

15. Date completed: _____

Figure 13.8
Metal Roof Condition

1. Building: _____ 2. Roof section #: _____

3. Date inspected: _____

4. Roof deck underside:

Overall condition	☐ good	☐ fair	☐ poor
Rusting or rot	☐ none	☐ minor	☐ major
Signs of leaks	☐ none	☐ minor	☐ major
Sagging	☐ none	☐ minor	☐ major
Uneven joints	☐ none	☐ minor	☐ major

5. Interior ceilings:

Water stains	☐ none	☐ minor	☐ major

6. Interior walls:

Water stains	☐ none	☐ minor	☐ major
Efflorescence	☐ none	☐ minor	☐ major

7. Metal roof surface:

Creased panels	☐ none	☐ minor	☐ major
Bent panels	☐ none	☐ minor	☐ major
Panel tears	☐ none	☐ minor	☐ major
Denting	☐ none	☐ minor	☐ major
Corrosion	☐ none	☐ minor	☐ major

8. Fasteners:

Missing	☐ none	☐ minor	☐ major
Corrosion	☐ none	☐ minor	☐ major
Loose	☐ none	☐ minor	☐ major

9. Flashing:

Overall condition	☐ good	☐ fair	☐ poor
Corrosion	☐ none	☐ minor	☐ major
Detached from wall	☐ none	☐ minor	☐ major
Open joints	☐ none	☐ minor	☐ major
Ridging	☐ none	☐ minor	☐ major

Figure 13.8
Metal Roof Condition *(continued)*

10. Roof penetrations:

Overall condition	☐ good	☐ fair	☐ poor
Flashing condition	☐ good	☐ fair	☐ poor
Properly sealed	☐ yes	☐ no	
Open seams	☐ good	☐ fair	☐ poor
Physical damage	☐ good	☐ fair	☐ poor

11. Comments: _____

_____

_____

_____

SHINGLE AND TILE ROOFS AND THEIR MAINTENANCE ACTIVITIES ────────────────────

Shingle and tile roofs have been widely used in applications having a roof pitch of at least three inches per foot. Most shingle and tile roofs use a supporting deck of wood or masonry, covered with a layer of asphalt-impregnated roofing paper. A wide variety of surface materials may be used based on the desired appearance and performance, including asphalt shingles, roll asphalt, wood shakes and shingles, asbestos, slate, clay tile, concrete tile, and aluminum tile. Materials are usually installed in a layered arrangement with staggered joints. The most commonly used fasteners include nails and metal clips.

One of the major advantages of shingle and tile roofs is their low requirement for maintenance. The relatively steep pitch to the roof quickly sheds water from the roof's surface, eliminating many of the problems associated with other, lower-pitch types of roofs. With the exception of broken and missing tiles, most leaks occur as the result of failures in flashings. Expected lives for these roofs range from 20 to 30 years for asphalt shingles, to more than 100 years for clay tile.

The most serious drawback for shingle and tile roofs for commercial, institutional, and industrial applications is that their steep pitch cannot readily support the installation of mechanical equipment. Exhaust fans, rooftop air handlers, cooling towers, and other mechanical equipment commonly found on other roof types must be located elsewhere.

Common Shingle and Tile Roof Problems

Although shingle and tile roofs are low in maintenance requirements, a number of items will require attention if the roofs are to perform effectively for their rated lives. The most common problems encountered include: (1) flashing failures; (2) curled shingles; (3) cracked shingles or tile; and (4) missing shingles or tile.

Flashing material corrodes and fails, and hence, many of the problems with shingle and tile roofs involve the flashing. Fasteners and mortar fail, allowing the flashing to pull away from vertical surfaces. Valley flashing is subject to damage from ice and hail. Repeated cycles of heating and cooling build thermal stresses which fatigue the metal, leading to cracks. All flashings should undergo a detailed inspection once a year.

Curling at the edges, or warf, is caused when exposure to heat dries out asphalt and wood shingles and shakes. Curling and warping are signs that the roof is reaching the end of its useful life and will need replacement in the near future. While moderate curling does not pose a risk of roof leaks, extensive curling indicates that the shingles are in danger of splitting or cracking, a condition that can lead to shingle loss and roof leaks.

Cracked shingles and tile are caused by overtightened fasteners or by physical damage to the roof. Physical damage may be caused by maintenance personnel walking on the roof, or by impact from overhanging tree limbs. Cracked shingles do not pose an immediate threat to the roof unless the cracks loosen the shingles and make them subject to lifting by wind. Cracked shingles can be secured through the use of roofing cement. Cracked tiles should be replaced as quickly as possible.

Missing shingles and tiles are most often caused by high winds, particularly on older roofs where the fasteners have become corroded and weakened. In some cases, physical damage to the roof loosens fasteners or cracks shingles and tiles at the fasteners, making them more susceptible to damage from the wind. When shingles and tiles are found missing, they should be replaced individually as quickly as possible to prevent the spread of the damage to adjoining units. If a large number of shingles or tiles are missing from a particular roof, it is an indication that the roofing materials or the fasteners are failing and the roof should be scheduled for replacement in the near future.

Shingle and Tile Roof Inventory

Complete an inventory of all shingle and tile roofing installed in the facility. An accurate inventory is critical to the establishment of a maintenance program for these roofs. Use Figure 13.9 to record inventory information for the shingle and tile roofs. A separate form should be completed for each section of a roof on a facility. In addition to the inventory form, a sketch should be completed for each roof section showing the outline of the roof and the location of all roof penetrations and flashings. When defects are found during a roof inspection, they should be noted on the roof to help in future problem diagnostics.

Shingle and Tile Roof Maintenance Activities

The most important maintenance activity for shingle and tile roofs is the completion of an inspection once a year, typically during the spring. As the roof ages and shingles begin to curl or crack, it may be necessary to complete an additional inspection of the roof in the fall in order to detect developing problems early. During each inspection, particular attention should be paid to roof flashings to ensure they are in good condition and properly attached. Use Figure 13.10 to record inspection information for shingle and tile roofs. A separate inspection form should be completed for each section of roof on a facility.

IMPLEMENTING A ROOF MAINTENANCE PROGRAM

Start-up efforts for a new roof maintenance program will be both labor and cost intensive. The roof inspections will have to be completed by maintenance personnel who are trained in roofing maintenance. Those inspections will identify a large number of repairs required, repairs that will also have to be completed by properly trained personnel. The high cost of implementation often causes organizations to defer the program, resulting in the continued decline of the roofing investment.

Figure 13.9
Shingle and Tile Roof Inventory

1. Building: _____ 2. Section #: _____

3. Area (sq ft): _____ 4. Pitch: _____

5. Year installed: _____

6. Tile manufacturer: _____ 7. Installer: _____

8. Deck:
   By site inspection, determine the deck material.

   ☐ gypsum slab   ☐ wood
   ☐ precast concrete   ☐ other _____

8. Shingles:
   By site inspection, determine the type of shingles installed.

   ☐ asbestos   ☐ cypress
   ☐ asphalt   ☐ mission tile
   ☐ cedar   ☐ pine
   ☐ clay tile   ☐ slate
   ☐ concrete tile slate ☐ other _____

9. Weight of shingles: _____

10. Flashing:
    By site inspection, determine the type of flashing installed on the roof. Check all that apply.

    ☐ aluminum   ☐ PVC
    ☐ copper   ☐ stainless steel
    ☐ galvanized steel   ☐ other _____

11. Comments: _____

_____

_____

_____

12. Date completed: _____

Figure 13.10
Shingle and Tile Roof Condition

1. Building: _____ 2. Roof section #: _____

3. Date inspected: _____

4. Roof deck underside:

Overall condition	☐	good	☐	fair	☐	poor
Rot	☐	none	☐	minor	☐	major
Signs of leaks	☐	none	☐	minor	☐	major
Sagging	☐	none	☐	minor	☐	major

5. Interior ceilings:

Water stains	☐	none	☐	minor	☐	major

6. Interior walls:

Water stains	☐	none	☐	minor	☐	major
Efflorescence	☐	none	☐	minor	☐	major

7. Shingles and tiles:

Overall condition	☐	good	☐	fair	☐	poor
Curled shingles	☐	none	☐	minor	☐	major
Cracked shingles	☐	none	☐	minor	☐	major
Missing shingles	☐	none	☐	minor	☐	major

8. Fasteners:

Overall condition	☐	good	☐	fair	☐	poor
Corrosion	☐	none	☐	minor	☐	major
Missing	☐	none	☐	minor	☐	major
Loose	☐	none	☐	minor	☐	major

9. Flashing:

Overall condition	☐	good	☐	fair	☐	poor
Corrosion	☐	none	☐	minor	☐	major
Detached	☐	none	☐	minor	☐	major
Open joints	☐	none	☐	minor	☐	major
Ridging	☐	none	☐	minor	☐	major

10. Comments: _____

_____

_____

If the area of the roofs that are to be added to the roof maintenance program is small, the program can be started as soon as the first inspection is completed, and information gathered on the roof's construction and history. However, if the facility has a large number of roofs that have not been well maintained, it will be necessary to phase in the program. Information will have to be gathered on the existing roofs. Inspections will have to be completed. Roofs will have to be tested. And while these activities are being completed, there will still be ongoing roof problems due to the lack of maintenance in the past.

The first roofs to be tested should include the oldest roofs in the facility and those that have a history of leaks and other problems. Implementing the program for these roofs will provide the greatest level of return in terms of energy and maintenance savings. ***Rule of Thumb:*** If a three-year testing cycle is implemented, approximately one-third of the roofs in the facility's inventory should be included in the first year's program. An additional third of the inventory should be added each year for the next two years.

GUIDELINES FOR DETERMINING WHETHER TO REPAIR, REPLACE, OR RECOVER A ROOF

Even with a comprehensive maintenance program, roofs will deteriorate with time and exposure. As they deteriorate, the need for maintenance will increase. Eventually, the point will be reached where it makes more sense to replace the roof than to maintain it. Unfortunately, the decision has traditionally been made more on an availability of funding basis rather than on need. If sufficient funds were available, the roof was replaced. If some funds were available, but not enough funds for replacement, the roof was recovered. If funds simply were not available, the roof was patched regardless of condition.

To be effective, repair, recover, and replacement decisions must be based on the condition of the existing roof. There are three key questions that must be considered. What types of problems have been experienced with the existing roof? What is the extent of the existing damage to the roof? Can repairs return the roof to an acceptable condition? Answers to these questions will help determine the best course of action.

While there are no hard rules as to when to repair, recover, or replace, there are several guidelines that can be followed.

☐ 1. Extent of water penetration. If more than 25 percent of the insulation is wet, the roof should be replaced. Recovering should not be considered an option with this amount of moisture in the insulation.

☐ 2. Age of roof. If the roof is less than five years old and the problems are isolated rather than throughout the roof, repairs should be made. If the roof is between five and ten years old, in nearly all cases it makes more sense to repair the existing roof than to replace it. At ten years old, the roof still has not reached its projected half-life, but differences between installations and the history of the particular roof may dictate that recovering or replacement be considered.

☐ 3. Condition of core samples. Regardless of age, if testing of the core samples indicate that the roofing material has lost sufficient tensile strength to be below the minimum recommended level by the roof's manufacturer, typically less than 100 psi, or if the samples show the material has become brittle, the roof should be scheduled for replacement. Embrittlement and loss of tensile strength are warning signs that the roof membrane is at risk of failing.

☐ 4. Costs. When the annual costs of maintaining a roof exceed 5 percent of the estimated replacement cost, excluding inspection and testing costs, consider replacing the roof.

☐ 5. Recover candidates. If the roof is approaching its design life and has not experienced major problems, and the roof moisture survey has shown that there is only a limited amount of hidden damage to the roof, the roof may be a candidate for recovering. Before the decision to recover can be made, a structural analysis must be completed of the roof deck and supporting structure to determine that it will be able to carry the additional load imposed by the roof recovering.

SUMMARY

Roofs are under constant attack from weathering forces, thermal stresses, and chemical action. Over time, these forces accelerate the normal aging process for the roof, eventually leading to roof failure. Most roofs on commercial, institutional, and industrial facilities do not fail suddenly. The failing process is a long and slow one that begins with small defects, defects that when left undetected and uncorrected evolve into major problems that waste energy, result in damage to the roof's underlying materials, cause damage to building finishes and contents, and shorten the life of the roof.

A maintenance program that begins with an inventory of all installed roofs in the facility and includes a routine of regular roof inspections and testing will extend the service life of every roof type while reducing energy losses through the roof. In most cases, maintenance managers have found that they have nearly doubled the average life of a roof by simply performing maintenance in a timely manner. Additional dividends are paid through lower maintenance costs, avoided interruption of services, and reduced energy costs.

Chapter 14

Foundation and Exterior Walls

This chapter presents information on how to maintain foundation and exterior walls to promote energy conservation. The information may be applied to all types of exterior wall and building foundation construction commonly found in commercial, institutional, and industrial facilities. For each different type of exterior wall construction, maintenance activities are identified that can be used in the establishment of a preventive maintenance program to reduce their maintenance requirements, maintain their appearance, lengthen their service life, and enhance their energy performance.

FOUNDATION AND EXTERIOR WALL ENERGY MAINTENANCE REQUIREMENTS

Exterior walls are a major component in a building's protection against the elements. In mid- and high-rise construction, they represent the facility's largest exposure. In low-rise construction, they are second only to the facility's roof.

Like the roof, exterior walls are subject to attack from ultraviolet light, heat, cold, rain, snow, hail, wind, pollution, wildlife, and people. They must resist these elements and present a waterproof seal, while maintaining the aesthetics of the facility. While exterior walls are considered to be low maintenance, they do require some maintenance if they are to perform these functions effectively.

Foundation walls are subject to attack from groundwater, lateral forces exerted by the ground, and vegetation. In addition to resisting these forces, they must provide adequate support to the facility while maintaining a waterproof seal. Foundation walls, like exterior walls, are low maintenance. Low maintenance, though, does not mean that they can be ignored.

Thermal Properties of Foundation and Exterior Walls

The rate at which heat is lost or gained through foundation and exterior walls depends on a number of factors, including the interior space temperature, the climate where the facility is located, and the wall's thermal conductivity. Two additional factors influence heat loss and gains for above-ground walls: the orientation of the wall, and the percentage of the sun's energy absorbed by the wall. Of these factors, the one that you have the most control over is the thermal conductivity of the wall.

The thermal conductivity of a wall rates its ability to conduct heat. Measured in Btu per hour per square foot of wall area per degree Fahrenheit temperature difference across the wall, thermal conductivity can be used to estimate the energy efficiency of a wall. The lower the thermal conductivity, the higher the energy efficiency. Figure 14.1 lists the thermal conductivity for a variety of wall types.

The U-values listed in the figure are for walls in good condition with no moisture present. The presence of moisture significantly changes the wall's overall U-value, resulting in an increase in heat losses and gains through the wall. The quantity of moisture present in the wall is directly related to the condition of the wall and how well it has been maintained.

The Impact of Neglected Maintenance

There are two major factors that contribute to the energy efficiency of exterior walls: moisture and infiltration. Moisture that penetrates the surface of the wall can migrate to the wall's insulation, reducing its effectiveness. Moisture also decreases the thermal resistance of the wall materials. For example, the thermal resistance of a brick wall decreases by approximately 16 percent with an increase in moisture content of only 5 to 10 percent. Moisture that gains access to the walls also leads to extensive damage to exterior wall materials, including pealing paint, the rotting of wood components, the corrosion of metal components, the failure of fasteners, and the spalling of masonry surfaces.

The second factor contributing to the inefficiency of exterior walls, infiltration, is the process by which outside air enters the facility through small cracks and voids in exterior walls. These cracks and voids increase the facility's heating and cooling loads in two ways. First, any air that enters the facility through the cracks and voids must be heated or cooled to the interior space temperature of the facility. Second, cracks and voids allow additional moisture to enter the facility, further damaging wall insulation and reducing energy efficiency.

The energy efficiency of foundation walls is impacted most by moisture. If the insulation is installed on the outside of the foundation wall, moisture can work its way between the insulation and the outside of the foundation wall, causing the insulation to separate from the wall. If the insulation is installed on the inside surface of the foundation wall, moisture migrating through the wall can saturate the insulation, destroying its effectiveness.

The key to energy efficient exterior and foundation walls is keeping moisture from penetrating the wall material and saturating the wall insulation. Keeping moisture out requires an ongoing program of inspection and maintenance. The specific inspection and maintenance activities will vary with the type of construction.

Figure 14.1
Foundation and Exterior Wall Thermal Conductivity

Foundations	U-Value (Btu/hr-sq ft-^{o}F)			
	8 " Block	12" Block	8" Concrete	12" Concrete
Plain	0.52	0.47	0.25	0.18
1" insulation board	0.17	0.16	0.13	0.10
3 1/2" fiberglass & drywall	0.08	0.07	0.06	0.05
1" insulation board, 3 1/2" fiberglass & drywall	0.06	0.05	0.05	0.04

Wood Frame Wall	U-Value (Btu/hr-sq ft-^{o}F)		
	Wood Siding	Face Brick	Stucco
Plywood sheathing & drywall	0.38	0.46	0.51
1/2" foamboard & drywall	0.25	0.28	0.28
Plywood sheathing, 3/12" fiberglass & drywall	0.07	0.08	0.08
1/2" Foamboard, 3 1/2" fiberglass & drywall	0.06	0.06	0.06

Masonry Walls	U-Value (Btu/hr-sq ft-^{o}F)	
	8" Block	12" Block
Plain	0.34	0.32
Face brick	0.19	0.18
Face brick, 3/12" fiberglass & drywall	0.06	0.05

THREE TYPES OF EXTERIOR WALL CONSTRUCTION ─────────

There are three major classes of exterior wall construction: wood frame, masonry, and metal. Wood frame construction finishes include shingles, weatherboard siding, mineral products, plywood, and face brick. Masonry construction includes brick, block, concrete, stone, and structural tile. Metal construction includes aluminum, corrugated iron or steel, enamel, coated steel, and a variety of protected metals.

Wood Frame Construction

Exterior wall wood construction consists of wood framing members, sheathing, and an outer layer. In most construction, insulation is installed between the wood framing members. Sheathing may be plywood, particleboard, fiberboard, or foamboard. The outer layer of the wall may be face brick, wood, vinyl, mineral board, or aluminum. In nearly all types of wood frame construction, a waterproofing layer is added between the sheathing and the outer layer.

The most common problem experienced with frame construction is the entry of water through defects in the exposed surface or through defects at the junction with other building components. Liquid water can enter the wall through cracks and voids. Water vapor can migrate through surface materials. Both liquid water and water vapor will contribute to the rotting of wood materials and the corrosion of fasteners. If the water penetrates the waterproofing layer, it will enter the wall insulation, decreasing its insulating abilities. Presence of water in the insulation will also promote the deterioration of framing members.

In severe cases of leaking, a discoloration of interior finishes will be noticed. For lower rates of leakage, it is difficult to detect leaking. Painted exterior wood surfaces will blister or peal as the result of trapped moisture. Face brick will show staining, spalling, or a light coating of efflorescence. Unfortunately most other types of materials will undergo little or no change in appearance. The damage will remain hidden until it becomes extensive.

Masonry Construction

Exterior wall masonry construction includes a variety of wall types, including both solid and cavity walls. Exterior materials include brick, block, concrete, stone, or structural tile. Interior finishes include brick, block, plaster, or drywall. Cavity walls typically use a layer of insulation between the inner and outer layers. Additionally, cavities in block construction may be filled with loose-fill insulation or foam insulation inserts. Masonry construction offers the advantage of long service lives while requiring very little maintenance.

There are two major maintenance problems with masonry construction—water and structural movement. Water can enter the facility through a number of defects in the masonry wall, including open vertical joints, cracks, porosity of the masonry units and the mortar, and spalled surfaces on the masonry units. Once inside, the water enters the insulation, destroying its insulating capabilities. Eventually it may migrate to interior surfaces, resulting in damage to interior finishes.

Settlement of masonry walls results in the formation of stress cracks that typically follow mortar joint lines, occasionally extending across a masonry unit. These cracks create entry points for water, allowing it to migrate to all areas within the wall, including the insulation.

One of the warning signs that water is entering a masonry wall is efflorescence on the exterior surface. Efflorescence is the white powder or light color crystals that form on the exterior of masonry walls. As water moves through the masonry materials, it carries dissolved salts that are deposited on the surface of the wall as the water evaporates. These deposits give the wall a chalky white appearance. While efflorescence does not cause damage and is considered primarily to be an aesthetic problem, it is an indication that moisture has penetrated the wall.

Metal Construction

Metal construction is similar to wood frame construction, with the use of steel framing supports instead of wood. The framing is typically covered with metal cladding. Insulation is placed between sections of the framing steel. The outer metal cladding can be made from galvanized steel, epoxy coated steel, or aluminum.

The most common problems with metal wall construction include failed fasteners, corrosion, and water penetration. Fasteners can be overtightened during installation, reducing their holding power. Wind loads on the wall cladding can induce vibrations that will work fasteners loose or cause them to fail from fatigue. Water can corrode fasteners, causing them to fail. Once the fasteners have failed, the metal cladding will loosen, destroying the integrity of the wall.

Corrosion of metal cladding weakens the panels, leading to tears and creases. Tears allow water to enter the wall cavities. If the tears extend to the fasteners, the metal cladding will loosen and pull away from the framing, allowing even more water to enter. Any water that makes it way past the metal cladding will accelerate deterioration of the wall components and will destroy the effectiveness of the insulation. Water trapped behind the metal cladding that freezes will stress the panels and fasteners, resulting in larger gaps and higher rates of water infiltration. The net result is a decrease in the wall's U-value and an increase in heating and cooling costs.

TWO TYPES OF FOUNDATION WALL CONSTRUCTION

The two most common types of foundation construction are block and concrete. In both types, a waterproofing coating is applied to the exterior surface of the foundation. Rigid board insulation may be applied to the exterior surface before backfilling. To help remove water from the exterior of the foundation, drain tile is installed at the base of the foundation wall and connected to a sump.

There are two common problems with foundation walls—settlement, and failure of the waterproofing layer. Settlement cracks destroy the integrity of the foundation wall's waterproofing, allowing water to enter the facility and damage interior insulation and finishes. In extreme cases, settlement cracks can lead to the failure of the foundation wall.

Most damage to a foundation wall waterproofing is the result of improper installation or the use of the wrong type of backfill material. Additional damage can be caused by the growth of the roots of trees and shrubs located too close to the facility. Once damaged, water will gain access through the wall, wetting interior insulation. Proper grading of the ground surrounding the facility will help to divert water away from the foundation.

FORCES ACTING ON EXTERIOR WALLS AND FOUNDATIONS ⸺

Foundation and exterior walls must resist a number of different forces. Understanding what these forces are and how they impact the performance of foundation and exterior walls will assist you in establishing maintenance programs designed to enhance their performance and maintain their energy efficiency. There are three general classes of forces acting on foundation and exterior walls: (1) heat and radiation forces; (2) water penetration; and (3) physical forces.

Heat and Radiation Forces

Heat and radiation exert a stronger force on exterior walls than they do on foundation walls due to their exposure to greater swings in temperature and their exposure to the sun. Foundation walls, however, are subject to some of the same forces, particularly those parts of building foundation walls that are above grade or those that are exposed to freezing ground temperatures.

Seasonal and daily temperature variations subject exterior wall and exposed foundation walls to thermal movement. During summer months, wall components exposed to direct heating from the sun often reach temperatures as high as 120 to 130 degrees Fahrenheit. Depending on the climate, these same materials may be exposed to temperatures during winter months that are well below freezing. The constant cycles of thermal expansion and contraction subject the materials to stresses, which often lead to fatigue failures. Over time, the stresses and fatigue failures result in the formation of hairline cracks in the materials. With repeated cycles of heating and cooling, these hairline cracks grow, allowing water to penetrate the surface of the wall materials.

Heat and radiation forces are particularly detrimental to building sealants used to fill the gaps between different materials and components used in construction of the foundation and exterior walls. The repeated cycles of heating and cooling put stress on the sealants and, when coupled with the effects of ultraviolet radiation from the sun, cause the sealants to lose adhesion. Eventually, the sealant fails and pulls away from one of the surfaces leaving a gap that allows water to penetrate the wall's outer surface.

Thermal stresses cannot be eliminated from foundation and exterior walls, but they can be reduced in some applications. The use of light colors on all painted exterior wall surfaces will reflect a portion of the sunlight striking the wall, reducing its temperature. Making certain that the proper type of sealants are used around openings in the wall will help to ensure that they perform properly over their expected life.

Water Penetration

Water penetration through cracks and gaps in exterior wall surfaces accelerates the deterioration of wall materials. Liquid water will cause permanent damage to many of the materials used in exterior wall construction, including steel and wood. Liquid water will promote the corrosion of steel wall components, and the decay of wood components. It will accelerate the deterioration of the finishes on both the interior and exterior surfaces of the wall.

Water in the form of vapor can enter foundation and exterior walls through small cracks and voids. As it moves through the interior of the wall, it condenses. Once condensed, it saturates wall materials, including insulation. If the wall is subjected to freezing, the condensed water will expand as it freezes, further opening cracks and gaps in the outer surface. Freezing water in concrete, block, and brick will cause spalling on the material's surface.

The major energy conservation concern is if the water gains access to the wall insulation, it will destroy much of its insulating value. Even a small quantity of water will impact the overall U-value of the wall.

Physical Forces

Physical forces acting on foundation and exterior walls produce both visible and hidden damage that can impact their performance and integrity. Physical forces that act on foundation and exterior walls include wind, structural movement, and plants. In most cases, the damage develops so slowly that it will go undetected without regular and thorough inspections.

Wind forces are more detrimental to wood walls than they are to masonry ones. Wind can cause sections of siding to loosen, making them even more susceptible to damage from wind. Loose siding also increases the chances of water entering the interior areas of the walls.

Structural movement, caused by uneven settlement of the facility, typically results in the formation of cracks in foundations and masonry walls. In severe cases, settlement can cause separation of the exterior wall from the foundation, providing a path for water to enter the facility, increasing air infiltration rates, and giving animals an easy route into the facility.

Plants impact both foundation and exterior walls. If trees are planted too close to buildings, their root systems can cause cracking of both. Trees and shrubs that are planted too close to exterior walls can cause damage as they come in contact with the walls, particularly on windy days. In both cases, the result is an increase in the potential for water to enter the walls and the facility.

Leaves from trees that overhang the facility or are planted too close can clog gutters and downspouts that normally would carry rain water well away from foundation and exterior walls. The result is an increase in groundwater adjacent to the facility, increasing the chances that the wall will leak. Leaking or overflowing gutters and downspouts are a leading cause of damage to exterior walls.

The best way to detect and minimize physical damage to foundation and exterior walls is to conduct regular and detailed inspections of the walls and the areas adjacent to the walls.

HOW TO DEVELOP THE FOUNDATION AND EXTERIOR WALL MAINTENANCE PROGRAM

The goal of a maintenance program for foundation and exterior walls is to implement activities that will minimize life-cycle costs while maintaining the appearance of the walls and minimizing energy losses. Achieving this goal requires that you first understand what types of foundation and exterior walls are installed in your facility. Once the inventory has been completed, an inspection program can be established that will identify problems as they are developing, well before they could lead to major damage to the existing structures.

Step 1: Develop the Foundation and Exterior Wall Inventory

The first step in the maintenance program is the completion of an inventory of foundation and exterior wall construction. Completing the inventory accomplishes two purposes. First, it identifies the type of construction used in the facility, a requirement in order to develop inspection activities. Second, developing an inventory of foundation and exterior walls in all buildings will help to ensure that all walls are included in the inspection program.

Use Figure 14.2 to develop the foundation wall inventory, and Figure 14.3 to develop the exterior wall inventory. If the same type of foundation or exterior wall is used throughout the building, only one form is needed for each. If the foundation or exterior wall construction changes in different areas within the building, use a separate form for each type used in the facility.

Step 2: Inspect Foundation and Exterior Walls

The major activity to be performed in a maintenance program for foundation and exterior walls is regularly scheduled inspections. The goal of the inspection program is to identify minor problems before they have had the time to develop into larger, more expensive ones.

For most facilities, inspections can be performed once a year. More frequent inspections may be required if a particular building has ongoing or developing problems, particular with water entering the facility. The inspection process should start

Figure 14.2
Foundation Wall Inventory

1. Building: _____   2. Location: _____

3. Area (sq ft): _____   4. Year installed: _____

5. Wall construction:
   - ☐ 8" block   ☐ 12" block
   - ☐ concrete: thickness (in): _____
   - ☐ other: _____

6. Exterior insulation:
   - ☐ none   ☐ 1" foamboard
   - ☐ 2" foamboard   ☐ other: _____

7. Interior insulation:
   - ☐ none   ☐ 3 1/2" fiberglass
   - ☐ 6" fiberglass   ☐ other: _____

8. Interior finish:
   - ☐ none   ☐ drywall
   - ☐ plaster   ☐ other: _____

9. Comments:

_____

_____

_____

_____

Figure 14.3
Exterior Wall Inventory

1. Building: _____ 2. Location: _____

3. Area (sq ft.): _____ 4. Year installed: _____

5. Type of construction:
   - ☐ wood frame   ☐ masonry
   - ☐ metal   ☐ other: _____

6. Exterior finish materials:
   - ☐ 8" block   ☐ 12" block
   - ☐ poured concrete   ☐ stucco
   - ☐ metal   ☐ wood siding
   - ☐ wood shingles   ☐ vinyl siding
   - ☐ aluminum siding   ☐ steel siding
   - ☐ asbestos-cement shingles   ☐ other: _____

7. Sheathing:
   - ☐ none   ☐ plywood
   - ☐ foamboard   ☐ other: _____

8. Insulation:
   - ☐ none   ☐ 3 1/2" fiberglass
   - ☐ 6" fiberglass   ☐ foamboard, thickness: _____
   - ☐ other: _____

9. Interior finish materials:
   - ☐ none   ☐ drywall
   - ☐ plaster   ☐ other: _____

10. Comments:

_____

_____

_____

_____

inside the facility, identifying areas where water stains or other damage is apparent. When inspecting the exterior of the wall, check the location and condition of all trees and shrubs growing close to the wall. One of the major causes of damage to exterior walls comes from portions of plants rubbing or falling against them. Additionally, the surrounding area should be checked for proper grading.

Use Figure 14.4 for foundation wall inspections. Figure 14.5 lists recommended inspection activities for wood frame walls, Figure 14.6 for masonry walls, and Figure 14.7 for metal walls. If the same type of construction is used throughout the building, only one form is needed for each building. If the exterior wall construction changes in different areas of the building, use a separate form for each type used in the facility.

Step 3: Test for Damage

If the inspection program identifies a number of areas where it is suspected that water has penetrated the wall and damaged the insulation, it will be necessary to determine the extent of the damage. One method that can be used to evaluate the condition of a wall's insulation is the use of a thermographic scanner.

Thermographic scanners detect the infrared energy emitted by the surface of objects. By detecting differences in both the wavelength and the intensity of the infrared energy emitted, the scanner can provide estimates of the rate of heat transfer through exterior walls. Areas with damaged or missing insulation will show up on the scanner's display as being warmer than areas where the insulation is intact.

In order to provide an accurate picture of the heat flow through the wall, the thermographic scan should be conducted only under certain conditions. The temperature difference between the building's interior and the outside air temperature should be at least 20 degrees Fahrenheit. There should be no wind, as wind blowing across the wall's surface will carry heat away, providing false readings. The scan should be taken early in the morning before sunrise to eliminate all effects of the sun as well as the thermal flywheel effect of the wall. The exterior surface of the wall should be shielded from heat sources, such as building exhausts and lighting. All interior lights, particularly those designed to flood the wall with light, should be turned off at least one hour before scanning the wall. If possible, all perimeter heat sources should also be turned off at least one hour before scanning the wall.

IMPLEMENTING A FOUNDATION AND EXTERIOR WALL MAINTENANCE PROGRAM

Initiating the inspection program for a facility where there is only a small number of buildings is not labor intensive or costly. However, if there is a large number of buildings, or if the buildings are complex, it will take time to initiate the program. If the facility is very large, it is recommended that the program be phased in over a several-year period, depending on the availability of trained inspectors. Phasing implementation will also help to spread the cost of making the repairs identified by the

Figure 14.4
Foundation Wall Condition

1. Building: _____ 2. Location: _____

3. Date inspected: _____

4. Outside conditions:
   Proper ground slope   ☐ yes   ☐ no
   Gutters & downspouts working properly   ☐ yes   ☐ no
   Trees proper distance away   ☐ yes   ☐ no

5. Structural wall
   Overall condition   ☐ good   ☐ fair   ☐ poor
   Water stains   ☐ none   ☐ minor   ☐ major
   Cracks   ☐ none   ☐ minor   ☐ major
   Leaking water   ☐ none   ☐ minor   ☐ major
   Efflorescence   ☐ none   ☐ minor   ☐ major

6. Interior finishes:
   Overall condition   ☐ good   ☐ fair   ☐ poor
   Water stains   ☐ none   ☐ minor   ☐ major
   Leaking water   ☐ none   ☐ minor   ☐ major

7. Comments:

_____

_____

_____

_____

Figure 14.5
Exterior Wood Frame Wall Condition

1. Building: _____ 2. Location: _____

3. Date inspected: _____

4. Outside conditions:
   Proper ground slope   ☐ yes   ☐ no
   Gutters & downspouts working properly   ☐ yes   ☐ no
   Trees & shrubs proper distance away   ☐ yes   ☐ no

5. Exterior finish:
   Overall condition   ☐ good   ☐ fair   ☐ poor
   Water stains   ☐ none   ☐ minor   ☐ major
   Peeling paint   ☐ none   ☐ minor   ☐ major   ☐ n/a
   Blistering paint   ☐ none   ☐ minor   ☐ major   ☐ n/a
   Rotted wood   ☐ none   ☐ minor   ☐ major
   Loose boards/siding   ☐ none   ☐ minor   ☐ major   ☐ n/a
   Corroded fasteners   ☐ none   ☐ minor   ☐ major   ☐ n/a
   Cracks & voids   ☐ none   ☐ minor   ☐ major   ☐ n/a
   Spalling bricks   ☐ none   ☐ minor   ☐ major   ☐ n/a

6. Interior finish:
   Overall condition   ☐ good   ☐ fair   ☐ poor
   Water stains   ☐ none   ☐ minor   ☐ major
   Peeling paint   ☐ none   ☐ minor   ☐ major   ☐ n/a
   Blistering paint   ☐ none   ☐ minor   ☐ major   ☐ n/a
   Rotted wood   ☐ none   ☐ minor   ☐ major   ☐ n/a

7. Comments:

_____

_____

_____

_____

Figure 14.6
Exterior Masonry Wall Condition

1. Building: _____ 2. Location: _____

3. Date inspected: _____

4. Outside conditions:
   Proper ground slope   ☐ yes   ☐ no
   Gutters & downspouts working properly   ☐ yes   ☐ no
   Trees & shrubs proper distance away   ☐ yes   ☐ no

5. Exterior finish:
   Overall condition   ☐ good   ☐ fair   ☐ poor
   Water stains   ☐ none   ☐ minor   ☐ major
   Pealing paint   ☐ none   ☐ minor   ☐ major   ☐ n/a
   Blistering paint   ☐ none   ☐ minor   ☐ major   ☐ n/a
   Spalling masonry   ☐ none   ☐ minor   ☐ major   ☐ n/a
   Cracks & voids   ☐ none   ☐ minor   ☐ major   ☐ n/a

6. Interior finish:
   Overall condition   ☐ good   ☐ fair   ☐ poor
   Water stains   ☐ none   ☐ minor   ☐ major
   Pealing paint   ☐ none   ☐ minor   ☐ major   ☐ n/a
   Blistering paint   ☐ none   ☐ minor   ☐ major   ☐ n/a
   Rotted wood   ☐ none   ☐ minor   ☐ major   ☐ n/a

7. Comments:

_____

_____

_____

_____

Figure 14.7
Exterior Metal Wall Condition

1. Building: _____   2. Location: _____

3. Date inspected: _____

4. Outside conditions:
   Proper ground slope   ☐ yes   ☐ no
   Gutters & downspouts working properly   ☐ yes   ☐ no
   Trees & shrubs proper distance away   ☐ yes   ☐ no

5. Exterior finish:

Overall condition	☐ good	☐ fair	☐ poor
Water stains	☐ none	☐ minor	☐ major
Peeling paint	☐ none	☐ minor	☐ major	☐ n/a
Blistering paint	☐ none	☐ minor	☐ major	☐ n/a
Corroded panels	☐ none	☐ minor	☐ major
Corroded fasteners	☐ none	☐ minor	☐ major
Loose panels	☐ none	☐ minor	☐ major
Cracked panels	☐ none	☐ minor	☐ major
Leaking seams	☐ none	☐ minor	☐ major

6. Interior finish:

Overall condition	☐ good	☐ fair	☐ poor
Water stains	☐ none	☐ minor	☐ major
Peeling paint	☐ none	☐ minor	☐ major	☐ n/a
Blistering paint	☐ none	☐ minor	☐ major	☐ n/a
Corroded panels	☐ none	☐ minor	☐ major	☐ n/a
Rotted wood	☐ none	☐ minor	☐ major	☐ n/a

7. Comments:

_____

_____

_____

_____

inspection program. If the program is to be phased, start with the oldest facilities or the ones with known problems first. These are the facilities that will provide the greatest rate of return on the program investment.

SUMMARY

Water is the biggest enemy of foundation and exterior walls. Water that has penetrated the wall's protection causes paint finishes to peal and blister, rots wood components, corrodes metal components and fasteners, damages the surface of masonry components, destroys the thermal resistance of insulation, and damages interior finishes. Undetected and uncorrected, it will lead to the need for very expensive repairs.

The best way to protect the integrity of foundation and exterior walls, and to ensure their thermal efficiency, is through the implementation of a preventive maintenance program that stresses ongoing inspections. Completed on an annual basis, these inspections will help to ensure that minor problems are detected and corrected before they can cause serious damage.

Section 5

How to Establish a Comprehensive
and Cost-Effective Energy
Maintenance Program

Chapter 15

How to Sell the Energy Maintenance Program to Senior Management

Nearly all organizations today have some form of a limited preventive maintenance program designed to reduce equipment downtime and to prolong the service life of select pieces of equipment and systems. Few organizations, though, have a fully comprehensive preventive maintenance program. Comprehensive programs are generally viewed by management as being too costly. Fewer still have implemented maintenance programs that stress energy conservation.

If comprehensive preventive maintenance programs are considered by the organization to be too costly, you face a serious uphill battle if you want to implement maintenance practices to save energy. Compounding the problem is the widely held belief that energy can be conserved only through the implementation of expensive, one-time projects. This ongoing search for the magic silver bullet of energy management has blinded many to the fact that real energy management is achieved only through an attention to detail that focuses on both day-to-day operations and long-term performance improvements.

If you are to be successful in gaining support for your efforts to reduce energy use through maintenance programs, you must educate the organization as to how energy use can be managed, then clearly demonstrate how your proposed maintenance program will help to achieve improvements in operating efficiency that will not only decrease energy costs, but also increase equipment reliability, decrease the frequency of breakdowns, and lengthen the service life of building systems and components.

This chapter presents information on the most important step in implementing a maintenance program to reduce energy expenses: winning support and approval of senior management. Regardless of the size and complexity of the facility, the program will temporarily increase maintenance expenses, particularly during start-up. Without this approval and support, management will balk at the perceived costs and shut down the program before it has had sufficient time to have an impact on equipment reliability, total maintenance costs, and energy costs.

The benefits of implementing a maintenance program for energy management may seem to be straightforward to the maintenance manager, but to the managers of the organization, things are not so clear. They see more money being spent with no return. Energy management through maintenance is viewed as being high-risk. After all, if the program could do all that its proponents claim, why haven't other organizations already implemented it? Remember how maintenance is viewed outside of the maintenance organization: a drain on resources. It is the first area to be cut, and the last to be restored.

The problem you face is that the organization has grown accustomed to playing the odds that equipment and systems will not fail if maintenance is deferred or ignored. It is an all-or-nothing gamble for them. Either things break or they don't. If they break, maintenance fixes or replaces them. If they don't break, they did not need the maintenance anyway.

To be successful at winning the support for the initiative, you must change this belief. You must make senior management realize that just because a piece of equipment or a system has not broken down, there is no guarantee that it is working efficiently.

Before trying to implement the program, you must take on the role of promoter. In this role, there are two tasks that must be accomplished. First, you should present information to the organization's senior management that makes the case for the program and overcomes organizational resistance. And second, the information should be sufficiently clear and strong to gain a commitment from senior management that not only supports the program, but also provides the necessary resources to implement and sustain it. Without the commitment, there will be no resources. Without the information, there will be no commitment.

A RECOMMENDED APPROACH FOR OBTAINING APPROVAL OF THE ENERGY MAINTENANCE PROGRAM

Any request for increased maintenance spending is met with resistance by senior management. Even in good times there is fierce competition for resources. Other units and departments within the organization are identifying opportunities to expand markets, improve services, or reduce expenses, all of which will improve profitability. Maintenance departments, on the other hand, have traditionally simply requested funds for "needed" projects and programs. While the need for the funding may be obvious to the maintenance department, most have failed to convince senior management of the need simply because they have not spoken in terms that senior management understands. Others speak of return on investment or paybacks. Maintenance managers typically speak of needs and wants, and possible equipment breakdowns. Needs and wants and possible breakdowns are not easily related to without supporting data. Perhaps that is why so few organizations consider maintenance spending to be an investment.

For example, many chillers limp along well past their useful service life because there is a failure to convince senior management that the chiller needs replacement. Terms like unreliable operation, frequent breakdowns, difficulty in obtaining replacement parts have no meaning or urgency to them. After all, the chiller is still running.

If you take this same approach in requesting funding for your maintenance program for energy management, you will have no chance for success. Remember, you are competing for funds with others who have done their homework and have presented their funding requests in terms that senior management understands. Even those seeking funding for energy conservation projects speak this same language. If you are to be successful, you also must speak the language that senior managers understand by presenting the benefits that the program will achieve.

Speaking the Language

A maintenance program for energy management must be promoted on the basis of its being a tool to reduce operating costs. To be successful, you should present hard data on the impact of maintenance and the lack of maintenance. After all, you are requesting that additional money be spent for maintenance when those who control the funding have little or no understanding of how spending funds on maintenance will benefit the bottom line. Remember, the popular view is that maintenance spending is a bottomless pit into which funds are poured and no return is received. It is unreasonable to expect financial managers to learn the language of maintenance. It is essential that you learn to communicate in the language of those who understand finance. You should learn and effectively apply cost analysis in selling your programs.

Start by looking at your current way of performing maintenance:

- What percentage of your available man-hours are spent performing emergency or breakdown maintenance?
- What percentage is spent performing planned or preventive maintenance?
- How many hours of overtime were required during the past year for emergency or breakdown maintenance?
- How much did those overtime hours cost you?
- How long is your deferred maintenance list, and how many man-hours would be required to complete it? Is it increasing or decreasing?

If you are going to build the case for spending more money on maintenance, even on a temporary basis, you must first know what you are currently spending and how you are spending it.

Next, you need to show where your operation stands in relation to those found in other organizations. Most organizations, unless they have a comprehensive preventive maintenance program, commit 80 percent or more of their labor resources to performing emergency and breakdown maintenance. In contrast, those that have a comprehensive preventive maintenance program commit less than 30 percent of their labor resources to emergency and breakdown maintenance. Show where your organization falls with respect to other organizations. The higher the percentage of time that you spend on emergency and breakdown maintenance, the greater the potential for overall maintenance and energy savings through implementation of the maintenance program.

How much the total maintenance costs will be reduced in the long term depends on your current allocation of resources to preventive and breakdown maintenance. ***Rule of Thumb:*** You can count on a 50 to 70 percent reduction in emergency and breakdown maintenance costs in a one-to-two-year period as a result of changing your operation to one that focuses on preventive and planned maintenance, as is the case with the maintenance program for energy management. This means that at an absolute minimum, you will experience a reduction in overtime requirements of 50 percent.

Since the focus of the maintenance program is energy conservation, you will need to determine and show the potential for energy savings that can be achieved as a result of the program's implementation. There are two methods for developing data to estimate and demonstrate the potential energy savings that can be achieved: (1) comparisons with other facilities; and (2) completion of a pilot program involving a key building system or component.

Perform Energy Comparisons

When evaluating the energy efficiency of a facility became a concern during the 1970s, facility managers looked to find a common yardstick by which they could measure their facility's energy use and compare it to that of other facilities. Although a number of different measures were developed and tried, the one that has received the most widespread acceptance is the energy use index (EUI). Expressed in Btu per square foot per year, the EUI allows facility managers to compare energy use in their facility to the use in other similar facilities in order to evaluate their relative efficiency.

Since the adoption of the EUI as a measure of energy efficiency, data have been compiled and published for a wide range of building types and designs located throughout the United States. One excellent source for this information is the document published by the Department of Energy, *Commercial Buildings Energy Consumption and Expenditures 1992*. By comparing the energy use of your facility to those in the published data, you can get a first approximation of your facility's relative energy efficiency.

There are some cautions that should be noted if EUI data are to be used to support the case for the maintenance program. EUI data do not correct for variations in climate. The data published represent average energy use across a wide range of climates. Since identical buildings located in two different climates will have different energy use rates, you must factor climate into any comparisons made between your facility and the published data.

EUI values do not take into consideration variations in use between facilities. For example, the EUI data do not differentiate between facilities of the same type and use that operate on different schedules. Some may operate for only 8 hours per day, while others may operate for 16 or 24 hours per day. Different operating hours will have a large impact on energy use. The EUI values, however, are simply averages for all facilities of that type and use. Before comparing the EUI for your facility to published EUI values, you must take into consideration use factors in your facility that will increase or decrease your energy use in comparison to that in other facilities.

Finally, the EUI does not address how well facilities are maintained or their overall condition. Again, the published data is a simple average of all facilities surveyed and includes a wide range of levels of maintenance.

In spite of these limitations, the EUI can help you build your case by showing the potential for energy cost reductions. It is particularly useful if you have access to energy use data from similar facilities having the same climate. If those facilities already have a comprehensive maintenance program in place, the data can be used to show the direct benefit that maintenance has on energy efficiency.

A more detailed discussion of energy use indexes and other methods of measuring energy performance is presented in Chapter 18.

Implement a Pilot Program

One of the most effective ways of convincing senior managers of the value of the proposed maintenance program is through the completion of a pilot project. Properly documented pilot programs will clearly demonstrate the value of the program in terms of cost savings to the organization and through improved system and equipment reliability.

The ideal pilot project is one that will have high visibility, a rapid and significant payback, and a high probability of success. To focus attention on the proposed program, the pilot project must capture the attention of senior managers. To do this, it must involve a system or a component that has a significant impact on facility energy use. The improvement that results from the maintenance activities performed must be easily documented and understood, such as through the use of meters. Too many guesses and assumptions about performance before or after weakens the case for the program.

Having a rapid payback is not enough. A project must have a rapid payback and must have a significant impact on facility energy costs. A project that saves $100 per year in energy costs, and costs less than $50 to implement is too small to generate interest and support for the program. While it may be worthwhile, it is not dramatic enough to serve as a pilot project. Pilot projects must be large in terms of their potential savings, and they must be significant. Although larger projects will require a larger investment, their greater impact on facility energy use is what is needed to demonstrate the value of the maintenance program.

The pilot project must also be one that has a high potential for success. Pilot programs can be risky, and the results may not always be exactly what was expected. If the existing condition is assumed to be worse than it actually is, and the projected results of the maintenance activity are greater than what is actually achieved, the energy savings produced will be less than what was anticipated. Therefore, it is very important that the project selected for the pilot program carry with it a high potential of success. Essentially, in selecting a project, you should stack the deck in favor of supporting the maintenance program.

Evaluating the potential for success for a particular project requires a detailed review of maintenance and energy use records for the system or equipment being considered as a candidate for the pilot project. Several factors should be considered in selecting the candidate. Ideally the unit should be in first one-third to one-half of its expected life. The unit should not have a history of major breakdowns or have ever undergone a major overhaul. Little or no preventive maintenance should have been performed on the unit. By screening units to meet these conditions, you will be selecting ones that will show the greatest improvement in performance through the completion of the required maintenance activities.

A Case Example

Although nearly all facilities will have a number of components and systems that meet these conditions, one component that offers the greatest potential for demonstrating energy savings through maintenance is the building's chiller. Building chillers are large energy users. A slight improvement in performance translates into large annual savings. If the chiller has been operating for several years, the chances that it is in good condition still are high. The potential for improving the performance of a chiller is large, particularly if preventive maintenance activities, such as cleaning heat exchanger surfaces, have been ignored. Many chillers also are separately metered, a requirement if the improvement in performance is to be measured and documented.

Review the maintenance records for the chiller, identifying all maintenance, including both routine and breakdown maintenance, that has been performed over the previous two-year period. Review the energy use records for the chiller, and determine the annual electrical energy use for each of the past two years. Test the full-load performance of the chiller to determine the kW required per ton of cooling produced. Using the information presented in Chapter 6, identify the recommended maintenance activities that should be performed on an annual basis for that chiller. Perform the recommended maintenance activities, tracking the total time and cost involved. Once the maintenance has been performed, repeat the full-load performance test of the chiller under conditions as close as possible to those that existed when the first performance test was completed. Compare the two test results to determine the decrease in kW per ton that is a result of the maintenance performed. Use this figure to calculate the annual savings that would result from the improved chiller performance.

Chillers are only one of a number of different systems that make ideal candidates for pilot programs. Other opportunities exist in all facilities. Lighting systems, boilers, large fan systems—all have the potential to serve as pilot projects for the program. No matter what system is used, the result of the pilot program must be documented and presented to senior management showing the cost and energy savings produced by the effort.

INFORMATION SENIOR MANAGEMENT NEEDS
PRIOR TO COMMITMENT

The single most important task that you face is gaining commitment from senior management to fund and support the program. In a climate where cutting maintenance costs is the accepted norm, it will take a very compelling argument to convince others that the program is worthwhile, justified, and cost-effective. Costs will increase initially as you will be adding maintenance and inspection tasks to workers' schedules, schedules that already are overcommitted in nearly all maintenance organizations. If the program is to get started in earnest, additional workers will be required. To simply try to add the new tasks to an already overloaded schedule will only ensure that the program will quickly stall and fail.

If you are to gain the commitment of senior managers, you must present a compelling argument for the program. You must show that while maintenance costs will increase in the short term, overall energy and maintenance costs will decrease. You must show what the current mode of operation is costing the organization in terms of increased energy use, inefficient use of maintenance personnel, decreased equipment life, and more frequent service interruptions. The key word in gaining commitment is *show*. Simply saying that the program is a good investment is insufficient. You must show what the program will accomplish.

In order to show how the program will benefit the organization, data will have to be gathered from a number of different areas. If a pilot program for the building's chiller or some other component has been completed, use it as an example of the impact that the maintenance program can have on energy use. Develop figures and graphs that clearly show the energy use of the chiller before and after performing the maintenance activities. Show what this improvement means in terms of the annual cost of running the chiller. Present the data on what it cost to implement the pilot program in terms of both labor and material costs. Calculate the simple payback period for the pilot project and show what this means to the organization over the life of the chiller.

Another source for data are the manufacturers of the energy using systems installed in the facility, such as boilers and cooling towers. Manufacturers can provide a reasonable estimate of the current operating efficiency of those systems based on their maintenance history, and what the operating efficiency would be under the proposed maintenance program. Again, develop figures and graphs to show what this improvement means in terms of energy and cost savings.

Although energy savings is a major driving force behind the program, it is not the only benefit that will be achieved. With the program in place, equipment operation will become more reliable, service interruptions will decrease, and equipment service lives will be extended. These effects must also be demonstrated to senior management. Again, simply stating that the program will make things better is insufficient; you need hard data.

Before showing how equipment operation will be improved, you must demonstrate the magnitude of the maintenance problem. Chances are, no one in senior management has any concept how many systems and components are installed in the facility and what is required to properly maintain them. While it would be too time consuming to develop a complete inventory of all systems and components, inventorying several key items will help to quantify the magnitude of the maintenance task. Square footage of roofs, number of air handler systems, number and capacity of building chillers, approximate number of lighting fixtures—providing hard data on what is currently installed in the facility will help nonmaintenance personnel to understand the magnitude of the installation. Briefly discuss what would be required of the program in order to properly maintain these items, and what the benefits of the program would be.

Select one item, such as building roofs, and go through its maintenance history in detail. Compare the actual service lives of roofs in the facility to what the industry considers to be the norm. Show what a lack of proper and comprehensive mainte-nance is doing to roof life and what it is costing the organization in terms of premature roof failures. ***Rule of Thumb:*** A comprehensive roof maintenance program will spend 5 percent of the replacement cost of the roof each year in maintenance, and will double the expected life of the roof. Show what you are currently spending on roof maintenance and what your facility's average roof life is. If proper maintenance extends the roof life to what the industry accepts as being the norm, show what the return would be for their investment.

Review the service records for the roofs, breaking down the labor and materials spent on preventive and corrective maintenance. Compare these costs to the costs of other similar organizations who already have a comprehensive maintenance program in place. Compare the frequency of leaks and emergency repairs experienced in your organization to what others have experienced. Show what roof leaks are costing the organization in terms of emergency repairs, damage to the facility and its contents, and interruptions of service.

Review the maintenance history for a critical piece of equipment, such as a building chiller or boiler. Identify the original cost of the equipment. For each time that the piece of equipment has broken down or required emergency service, identify the cause of the problem, what was required to correct it, how long the outage of services lasted, the cost of correcting it, and whether or not a maintenance program could have prevented it.

The danger is that you may receive a go-ahead to implement the program without receiving a commitment. Without that commitment, when costs start to increase, support will rapidly evaporate, resulting in shutting down the program before it has had a chance to produce positive results.

Sell the benefits of the program, but do not oversell them. Senior management must understand that it may take several years of effort before noticeable results are achieved. What you need is their backing and commitment to carry you through the time when maintenance costs have increased until the program is mature enough to start reducing both energy and total maintenance costs. At a minimum, you need a commitment to fund the program for two years.

HOW TO ENSURE ONGOING COMMITMENT
FOR YOUR PROGRAM

The effort to gain and maintain support for a maintenance program does not stop with the initial commitment. Conditions change. Priorities change. Without an ongoing effort on the part of the maintenance department to promote the new maintenance program and maintain support, enthusiasm and funding can soon be lost. Senior management must be kept informed as to the status of the program, and what progress is being made. Remember, there will still be equipment failures and interruptions of service. What you do not want happening is for others to assume that since these failures are still taking place, the program is not working.

The key to maintaining support and enthusiasm for the maintenance program is communication. You should take the initiative to publicize the program and to keep those who control the funding informed as to its progress and successes. Many different communication methods are available, including the use of newsletters, briefings, and memorandums. In most cases, you can make use of multiple methods to keep the organization informed on the status of the program and the progress that has been made. While self-promotion is generally not something that maintenance managers engage in, self-promotion is critical for sustaining the program, particularly if there are those in the organization who would rather commit the funding to other activities.

Presenting Status Reports

On a regular basis, you should report the status of the program to senior management, including what systems and building components have been inventoried and added to the inspection program. As inspections are completed, report the conditions found, including:

- Were there ongoing problems that negatively impacted performance and energy efficiency?
- What was required to correct them?
- What would the consequences have been if the problems had not been found and corrected?
- Did identifying and correcting the problems avert an interruption of services to a portion of the facility?
- How much money was saved by correcting the problems before they developed into larger ones?

These pieces of information must be reported regularly.

As the program is expanded to include certain energy using systems, such as boilers, chillers, and cooling towers, you should report hard data on before-and-after performance of those systems. Collect data on those systems both before and after performing the required maintenance. Calculate the annual energy savings resulting from the improved performance, and translate the energy savings into annual cost savings. Report what the maintenance effort cost to implement, and show the resulting payback for the effort. Reporting program successes and showing energy and maintenance costs that the organization has avoided will help to gain ongoing support while the program is evolving.

As the program matures, there will be a shift in the focus of the overall maintenance effort from emergency and breakdown maintenance tasks to scheduled and preventive activities. On a quarterly basis, give a breakdown of the number of hours spent on each type of maintenance activity. Discuss in the report what this shift in focus means to maintenance costs. Do not attempt to report more frequently than once per quarter to prevent having a single breakdown incident skew the data.

SUMMARY

An energy conservation program that achieves its savings through good maintenance practices is a cost-effective way of reducing energy costs while reducing overall maintenance costs. Unlike energy projects, it is a long-term effort that typically requires one or more years to fully implement. The benefit of maintaining equipment and systems at their peak operating efficiency typically pays for the initial cost of implementing the program in months, not years.

The single most important step in implementing the maintenance program for energy conservation is to gain the commitment and funding necessary to initiate and sustain the program. To gain that commitment, you must change the organization's view of maintenance, you must learn to speak the language of the financial manager, and you must show hard data that demonstrate the value of the maintenance program. An effective tool that will help to accomplish these requirements is a carefully selected pilot program.

Once the maintenance program has been implemented, you must work to keep attention focused on it by regularly communicating the program's achievements and successes. Without that focused attention, the funding that the program requires will fall easy prey to others who are looking to fund their pet programs.

If you are successful in gaining support and commitment for the program, and can put in place the necessary inspection and maintenance activities, the reward will be lower energy costs, lower overall maintenance costs, and longer equipment service lives.

Chapter 16

Guidelines for Planning
an Energy Maintenance Program
in Your Facility

All maintenance programs undergo a process of evolution. As the organization grows, so does the maintenance program. As conditions and priorities within the organization change, the maintenance program must change and adapt to the new requirements. Even when there is a need to drastically change an existing program, those changes are implemented gradually over time. Seldom does the maintenance manager suddenly replace an entire maintenance program with a new one. That same process of gradual evolution must be applied to any maintenance program that is being implemented to promote energy management.

The key to successfully modifying a maintenance operation or implementing a new maintenance initiative is the setting of priorities. The sheer size of most maintenance operations makes it difficult and risky to change everything at once. New maintenance initiatives and programs will require new information. Data will have to be collected on what systems and components exist in the facility. Maintenance personnel will have to develop and learn new maintenance procedures. You will have to learn, sometimes by trial and error, not only what maintenance tasks must be performed, but also when it is most appropriate to perform them, and how frequently they must be carried out.

This chapter presents information to assist you in setting priorities for implementing the maintenance program that will help to reduce energy use in a facility. Establishing these priorities is an essential step that must be completed before implementation of the program as discussed in Chapter 17. The information presented in this chapter may be used in any commercial, industrial, or institutional facility, regardless of its size or the speed with which the program is to be implemented. By following the steps suggested in this chapter, you will help ensure the success of your program by reducing the chances that the department will be overwhelmed by program start-up labor and funding requirements. Through careful planning, you will be able to minimize much of the confusion and road-blocking that often accompanies the implementation of a new program.

PLANNING THE IMPLEMENTATION

Once you have has gained the support and commitment required for the program, there is an urge to get the program up and running as quickly as possible. Upper management is watching and expecting to see action and results. If they do not see action or results immediately, the funding may be at risk and the program runs the risk of being terminated. Resist this urge. If you presented the program properly, senior management will understand that it requires time to initiate. They will understand that results will take even longer.

Still, though, the first impulse is to do everything at once. Enlist inspectors to start collecting the inventory data. Direct maintenance personnel to start performing the required maintenance activities. Record data coming back from the field and inform senior management of the progress being made. Constantly push to expand the program.

The problem with this approach is that it ignores one of the most basic requirements for successful implementation of new initiatives, planning. Remember, the basic mode of operation of the maintenance function is being changed. Reactive and breakdown maintenance is being largely replaced by planned and scheduled maintenance. No longer will it be sufficient to simply keep equipment running. Under the new maintenance program, equipment must be kept running efficiently. Priorities have changed, and new priorities dictate that maintenance personnel perform their work differently. To successfully accomplish this change, the approach that maintenance personnel take to their work must be changed first. Dumping new assignments on an existing workforce that has taken the same approach to maintenance for years will only serve to confuse and frustrate them, leading to an undermining of the new program. The following steps should be considered when starting a new program:

- Involve maintenance personnel. Clearly explain the reasons for implementing the changes and how the new program will operate. Tell them what the program is designed to accomplish and how it will impact the way in which they perform their duties. If a pilot program has been completed, show them what was done, why it was done, and what it accomplished. Project the results from the pilot program to other maintenance areas.

    Do not just talk to personnel; listen as well. Maintenance personnel are the ones closest to the operation and maintenance of the building's equipment and components. They know where the problems are, and most likely know what is required to correct them. They can help in identifying maintenance activities that must be performed under the new program, and they can assist in identifying priorities for the program. By making maintenance personnel part of the program planning, the maintenance manager is helping to ensure that they are committed to the new program and are working to make it their own.

- Determine tool and training requirement. Having maintenance personnel buy into the program should be part of the initial phase in planning the implementation. However, no matter how committed the workforce is, they will not be successful if they do not have the tools and training required to complete the new activities. In many cases, they will be performing maintenance activities that they have never performed before. While many of these activities are simple and straightforward, some are not and will require the purchase of expensive test equipment and the completion of training in how to use it properly. The purchase of tools and the completion of the training cannot be deferred. They are part of the program start-up costs and must be planned for.

  Tool and training requirements are not universal. The requirements depend in part on what maintenance activities have been performed in the past, and the type of new activities that are about to be implemented. For example, maintenance crews who are responsible for the efficient operation of the chillers in a facility will in most cases need additional tools, including clamp-on ammeters, megohmmeters, tube cleaners, and eddy current testers.

- Determine allocation of workforce. You must also plan the initial level of workforce commitment. Under the current mode of operation, the majority of the maintenance workforce has been performing breakdown and emergency maintenance tasks. Even if the new maintenance program were rapidly implemented, it would take time for it to take effect and reduce the frequency of equipment breakdowns. Therefore, committing a portion of the existing workforce to the new program will increase the workload of the remaining maintenance personnel who must continue to perform breakdown and emergency maintenance activities.

There is no set rule as to how to allocate the time of maintenance personnel during program start-up. Too small a workforce allocation to the new program will result in unacceptably long implementation times. Too large a workforce allocation will result in unacceptable delays in performing breakdown maintenance activities, and will increase the pressure to divert some of the dedicated workforce to perform breakdown maintenance activities. You must achieve a working balance that is acceptable to the organization and to the goals of the new maintenance program.

Start small with a commitment of no more than 10 to 15 percent of the maintenance workforce to the new program. Every month, review maintenance records to determine trends in performance. Is the maintenance department falling behind on breakdown maintenance? Has there been a change in the number of breakdowns occurring for those systems already in the new maintenance program? How many times were maintenance personnel pulled off their assignments under the new maintenance program in order to assist with breakdown maintenance? If breakdown maintenance is constantly detracting from the new maintenance program, you can either temporarily increase the size of the workforce, or slow the rate at which the new maintenance program is being implemented. It does no good to put a program in place that focuses on scheduled and preventive maintenance only to defer the maintenance activities. Deferred maintenance and energy management cannot co-exist in the same organization.

SETTING MAINTENANCE PRIORITIES

One of the biggest risks in beginning any new maintenance program is trying to do too much too quickly. The risk is even greater when one is trying to change the basic approach to maintenance from one where maintenance reacts to breakdowns and emergencies to one where maintenance emphasizes activities designed to prevent breakdowns and limit emergencies. While this change in approach is taking place, maintenance personnel will have to play catch-up for maintenance activities that have been ignored or deferred in the past. It will take time for the new program to reduce the frequency with which systems and components fail. During this time, personnel will have to perform the new maintenance tasks as well as respond to the breakdowns that will continue to occur.

If too many systems and components are immediately placed into the new program, you run the risk of overcommitting the workforce. Since breakdowns and emergencies cannot be ignored, the maintenance activities required by the new program will be deferred. Deferring those activities defeats the purpose of the program, ensuring that the past history of breakdowns will continue.

Attempting to start the program too widely also makes it difficult to modify the program to meet conditions found in the facility. There is no one correct way to run the maintenance program. The maintenance procedures and schedules presented in the preceding chapters are guidelines to be used in establishing the program. The existing conditions in a particular facility may require modifying one or both. Additional maintenance activities will be added. Maintenance frequencies will be changed to match the needs of the organization. While some modifications can be anticipated at implementation, many cannot and will be identified only through trial and error. Recognizing the need to make changes and implementing those changes becomes much more difficult if the program starts out by attempting to include a large number of items.

Another risk that you face in implementing the program is starting with the wrong systems or components. While it is important that the program include as many building systems and components as possible when it is fully implemented, the return, including both the energy and the maintenance savings, achieved by implementing the maintenance program for different building systems and components varies widely. Doubling the number of systems or components included in the program does not mean that the savings will also be doubled, particularly if the primary energy using systems are added first. There will be a diminishing rate of return as other, less energy intensive systems and building components are added to the program. Selecting items that offer a low rate of return will result in increased maintenance costs without rapidly realized offsetting energy and maintenance cost reductions, weakening support for the program.

The best way to avoid these and other related start-up problems is to begin the program slowly, phasing implementation as resources permit and experience is gained. The rate at which items are added to the program will vary with the size of the facility, the availability of maintenance resources, and the rate at which breakdowns and emergencies are currently being experienced.

To establish a phasing schedule, you must first:

- Know why the program is being implemented. What is the driving force behind the program? Is it energy conservation? Is improving equipment reliability the most important factor? How important is reducing overall maintenance costs? While the maintenance programs will reduce both energy and overall maintenance costs, and will improve the reliability of components and systems, the impact of adding various building systems and components will vary. How those items are phased into the program will change depending on the overall goals of the program.

- Know how many of what types of systems are installed in the facility. While it is generally assumed that managers know what is installed, most underestimate the actual figure. This is particularly the case in facilities that are constantly being modified or updated to meet the changing needs of building occupants. Unless there is an up-to-date inventory of systems and equipment installed in the facility, it will be difficult to estimate program start-up requirements.

- Estimate the labor requirements necessary in order to implement a certain phase of the program. Estimating the labor requirements necessitates understanding what maintenance tasks need to be performed and the frequency with which they must be performed. Additionally, since maintenance requirements may have been ignored in the past, there will be additional labor requirements to complete these tasks and bring the equipment and systems up to acceptable performance standards. Underestimating the labor requirements will result in delays in implementation or will force corners to be cut, an action that defeats the whole purpose of the program.

HOW TO ESTABLISH PROGRAM GOALS

Program goals help to establish the order in which building systems and components are added to the maintenance program. By establishing the program goals you give the program a focus. Goals tell you where you are going and why. They help to ensure that you do not lose sight of the purpose of the program and head off in unproductive directions. They help you to concentrate your resources where they will do the most good. Focused goals are particularly important when one realizes that maintenance managers are constantly being pulled in different directions by different groups in the organization. Goals may come from the maintenance department, or they may be established by senior management.

Goals also help you to gauge the program's performance over time. They give you a means of determining where the program is with respect to where you want it to be. Without having set goals, it is difficult to determine how well the program is meeting expectations.

Goals must be specific and measurable. If one of the program's goals is to achieve energy savings, the goal must clearly state the reduction target and how to determine if you have achieved it. Instead of stating that the goal of the program is to reduce energy use, set a specific value of how much you want the program to save, and when you want to achieve those savings. For example, a program that is being implemented over a four-year period may set goals of a reduction in energy use of 5,000 Btu per square foot for the first year, an additional 10,000 Btu per square foot for the second year, and an additional 7,000 Btu per square foot per year for each of the next two years.

Goals must also address the what and when of program development. What equipment or systems are going to be added to the program? When are they going to be added? Again, the goals must be specific and measurable. Instead of stating that the program will add all HVAC systems during the first and second years, spell out what systems are going to be added, how many of those systems there are, and exactly when those systems are going to be added. Set goals that are concrete. For example, one goal might be that the initial survey and required maintenance will be performed on ten centrifugal chillers and ten cooling towers during the first year of the program's operation. During the second year of operation, the program will survey and perform the required maintenance on five building boilers and forty air handling systems.

By stating the program goals in this manner, you can constantly check on the status of the program and progress being made to determine if additional resources must be committed. If the facility is particularly large or complex, the use of interim goals can help to keep the program on schedule by tracking accomplishments and performance on a monthly or quarterly basis.

Goals must also be attainable. Setting unrealistic goals will undermine support for the program. Maintenance personnel will find that they are falling behind. Maintenance activities will be rushed or deferred, just so that additional items can be listed as being included in the maintenance program. Senior management will falter in their commitment for the program as the promised results will not materialize. If the program advances more slowly than was projected, or if corners are cut and maintenance is deferred, the program will not achieve the savings and benefits that were promised.

Even when programs are carefully laid out, things do not always go as planned. Organizational plans change. Equipment breaks down. Demands on maintenance workers' time may cause the program to progress more slowly than originally planned. And with the program behind schedule, the anticipated benefits and savings will be delayed. Unless there can be a revision in the assignment of personnel, the program's goals will have to be revised to reflect the reality of the situation. In either case, senior management must be kept informed of developments and made part of the decision-making process. The last thing that senior management wants is a surprise. Surprises, such as reduced accomplishments, erode support for the program. By keeping them informed and making them part of the decision-making process, you will increase the chances of retaining their support and commitment, and may even be able to convince them to provide more support to accelerate implementation.

Step 1: Conduct a Complete Inventory

The first step in identifying components and systems to be included in various phases of the maintenance program is the identification of what is installed in the facility and where. Few facilities, unless they already have a comprehensive preventive maintenance program in place, have a complete inventory of building systems, components, and equipment. That inventory must be developed before it can be determined how the program is to be phased. Without a complete inventory, selecting individual systems and components will be a hit-or-miss process that depends on the memory of maintenance personnel.

Complete the inventory using the forms presented in this handbook. For the preliminary inventory, it is critical that all pieces of equipment and building components be identified well before the phasing schedule is established and before targets for energy and maintenance cost reductions are established. Components that are missed will result in program delays and increased expenses, both of which contribute to a loss of support for the program after it has been initiated. Underestimating the installed equipment base is a common and very serious error in establishing maintenance programs.

Step 2: Estimate Man-Hour Requirements

While the inventory is being completed, it will be necessary to determine approximate times required to perform the required maintenance on each type of system or component to be included in the program. These time estimates are important in determining what can be accomplished during a given period of time with a particular labor force. Work with maintenance personnel and equipment manufacturers to estimate the number of hours to perform the identified maintenance activities for a particular type of system or building component. What you are looking to determine is the average number of hours that will be spent on that item. Some will take more or less time, depending on their particular conditions. For estimating man-hour requirements though, an average time is all that is needed.

Multiply the number of man-hours required for a particular system or component by the number of those items that exist within the facility to determine the total man-hours required to implement the program for that system or component. When compared to the total man-hour availability from the maintenance force, this value will help you to identify how many systems can be implemented in a given period of time.

Step 3: Review Equipment Purchase Orders

In most facilities, particularly those that are large, the inventory will be out of date before it is completed. Equipment will be added. Systems will be upgraded or replaced.

Buildings will be remodeled. It is surprising how quickly the facility changes. Unless you establish procedures for modifying the inventory as these changes take place, the inventory will never be complete and up-to-date. Every building remodeling job, no matter how small, requires a survey when it is completed. All new equipment must be added to the system as it is installed. To help in identifying new and replacement equipment and components that are being installed, all equipment purchase orders should be reviewed. Purchase orders that involve the purchase of items that need to be added to the maintenance program inventory should be flagged for site inspection.

Step 4: Select Systems and Components

Maintenance programs for energy management can fail simply because the wrong building equipment and components were selected for implementation at program start-up. Financial managers are closely watching the program, looking for the savings to develop. They have been promised savings in energy and total maintenance costs. If you have properly sold the program, the financial planners understand that total maintenance savings will not be achieved until some time in the near future. They also expect to see energy savings, however slight, almost immediately. If the wrong building equipment and components are selected to initiate the maintenance program, they will not produce the savings that the financial managers are expecting. And if financial managers do not get the savings they are expecting, the program will be short-lived. Therefore, it is critical that you carefully select items that will provide the greatest possible return in the shortest possible time.

The items that will provide the largest return in terms of energy savings are the primary energy using systems in the facility: boilers, chillers, cooling towers, and other similar systems. It is with these systems that slight improvements in operating efficiency will translate into major annual energy savings. Review the inventory information for major energy using systems, and estimate the labor requirements in order to add them to the maintenance program.

Timing plays an important role in selecting what systems to start the program with. In order to perform the required maintenance activities, particularly if maintenance has been deferred on these systems in the past, they will have to be taken off line for several days. Not all facilities have the luxury of having backup boilers or chillers that will enable them to shut down one of the units for maintenance. If program start-up occurs during the heating season, consider performing the required maintenance activities on the chillers and cooling towers. If start-up occurs during the cooling season, start with the boilers. If both heating and cooling systems are required when the program is started, and backup systems are not available, consider starting with the lighting system. Lighting systems are major energy users and will provide a good rate of return through energy savings.

The energy savings that can be achieved through the performance of maintenance activities is only one factor that must be considered when selecting building equipment and components during program start-up. The role that other factors play in selecting items to include in the program varies depending on the needs of the organization.

One of the most significant of these factors to consider is maintenance costs. In addition to reducing the energy requirements of the facility, the maintenance program is designed to reduce total maintenance costs. Therefore, equipment and components that have proven to be high in maintenance costs under the current maintenance system are potential candidates for early inclusion in the program. Look back through the maintenance records for the facility. Are there particular items that stand out as requiring an unusually high amount of maintenance effort to keep them functioning? Is it just that one item, or do similar items require the same high level of attention? For example, if the maintenance records indicate that a particular single-ply roof has required a unusually high level of effort to stop leaks, it might simply be a case of a poor installation or that one roof needing replacement. But if the records indicate that all of the installed single-ply roofs require more maintenance effort than all other roofs, then single-ply roofs will be a good candidate for inclusion early in the program.

Another factor that will help to determine what items to include early in the program is how critical they are to the operation of the facility. If the item is a piece of mechanical or electrical equipment, is there a backup system that can be used in the event of the failure of the primary unit? For example, many facilities install central chillers in a parallel configuration. On most days, one chiller and one set of pumps can carry the building load. If the operating chiller or pumps fail, the backup chiller and pumps can be brought on line and supply chilled water to the required areas of the facility.

Not all mechanical or electrical systems have backups, nor do any of the building envelope components. But not having a backup is not sufficient reason to include the item in the initial phases of the maintenance program. The item must also be critical to the operation of the facility. The breakdown or failure of the item must result in a significant loss to the facility through disrupted operations or damage to the building or its contents. Identify those items that do not have backup systems and are at risk of causing major disruptions to operations or losses in the event of a failure.

When evaluating items to include in the initial phases of the maintenance program, consider their age and condition. Some equipment or components will be in such poor condition that it would not be cost-effective to include them in the maintenance program. In these cases, inclusion and performance of the required maintenance would not help anyway. The item simply may be in need of replacement. For example, a roof that is approaching the end of its rated service life, is leaking in a number of areas, and is showing signs of advanced deterioration should not be added to the maintenance program. Plans should be make to replace it as soon as possible.

Finally, there are the political considerations. The support of senior management is essential to the long-term success of the program. Everyone has their pet peeves when it comes to the operation and maintenance of the facility, including senior managers. It could be the appearance of a particular building, or the operation of an air conditioning system. Like it or not, if the pet peeves of senior management are not addressed by the program, support will be lost. And lost support may be the difference between long-term success and loss of funding. Talk to senior management and make them part of the decision-making process in setting priorities. While some of their suggestions and perceived needs may not require immediate implementation, it is surprising how often they can provide assistance in identifying items whose proper performance is important to the efficient operation of the facility.

SUMMARY

The implementation of an energy conservation program based on good maintenance practices requires careful planning. Too often maintenance managers try to rush implementation without fully understanding how the program will impact their maintenance operation. Rushing implementation results in false starts, missed program goals, and ultimately in program failure.

Careful planning, starting with involving maintenance personnel and senior management, will help to ensure that the program is properly directed at initiation. Proper tools and training must be provided if maintenance personnel are to be able to perform their required new duties. Maintenance time must be allocated to the program without causing a disruption of breakdown maintenance. An inventory of items to be included in the program must be completed. Goals for program performance must be established. Estimates must be completed on maintenance requirements.

To phase in the program, you should determine priorities for the program based on the needs of the organization, including energy conservation, reduced maintenance costs, reliable operation, and political considerations. Matching available manpower to program requirements based on the priorities of the organization will help ensure the success of the program.

Chapter 17

How to Implement an Energy Maintenance Program: Three Approaches

Maintenance programs for energy management, like the facilities they support, come in all sizes and shapes. Programs have been implemented in facilities ranging from as little as a few thousand square feet in a single building to multi-million square foot facilities scattered through hundreds of buildings over multiple sites. Programs have been implemented using in-house personnel, contract services, or a combination of both. Some of these programs have been fully manual in their operation while other have made extensive use of computerized maintenance management systems. Program start-up overhead costs have ranged from a few thousand dollars to several million.

How successful you are in implementing the energy maintenance program depends to a great extent on how well the program is matched to the requirements of the facility. Not all facilities have the ability to perform the required maintenance tasks in-house. Similarly, not all facilities will gain the same benefit from computerizing the operation. Some simply cannot afford the required investment. But lacking the in-house capabilities or the finances to invest in a computerized system cannot be allowed to stop the program from being implemented. There are alternatives. It is your job to find the one best suited to the facility.

This chapter focuses on the implementation options available to you and how to select the most appropriate one. The information presented in this chapter may be used in any commercial, industrial, or institutional facility of any size or complexity. By using the information from this chapter as a guide, you can develop a program implementation scheme that is best suited to that particular facility, a program that can effectively reduce energy and total maintenance costs.

There are three ways in which a maintenance program for energy management may be implemented: (1) using in-house maintenance personnel for all tasks; (2) contracting for all maintenance labor from outside maintenance firms; and (3) splitting the maintenance activities between in-house and outside maintenance firms. All three methods have been used effectively by maintenance organizations to perform break-down, planned, and preventive maintenance. All three can be used equally effectively in implementing an energy maintenance program.

HOW TO EFFECTIVELY USE IN-HOUSE PROGRAMS

Maintenance organizations have traditionally been operated primarily through the use of in-house personnel. When new initiatives, such as scheduled and preventive maintenance programs, have been implemented, new mechanics have been hired or maintenance personnel were reassigned from other duties. Staffing levels have been determined by the availability of qualified personnel and the ability to reassign them to the new duties. The same approach can be effectively used when establishing an energy maintenance program.

The Costs and Benefits of Using In-House Personnel

One of the most significant advantages of using existing personnel to operate the energy maintenance program is that they are already familiar with the facility and the equipment that must be maintained. Productivity will not suffer as a result of lost time spent searching for equipment and systems, and in some cases, buildings. Not only do in-house personnel know where the systems are, they know what areas of the facility are impacted by their operation. In-house personnel are familiar with the procedures that dictate how the maintenance department performs work. Institutional knowledge, an important factor in reducing program start-up costs, is simply not available to contract maintenance workers when the program is contracted out.

Another important advantage of using in-house personnel is the knowledge and experience they gain as the program evolves. When a maintenance program is staffed by in-house personnel, the workforce is relatively stable from year to year. Experience they develop in performing the maintenance tasks one year will assist in performing those same activities in following years. As they work with the equipment and components, they will recognize activities that are effective and ones that are not. They will recognize the need to perform additional maintenance tasks or the need to perform tasks more frequently. This recognition can develop only with experience working with those particular systems and components over time. When the program is contracted out, there is no guarantee that the same personnel will perform the tasks in subsequent years.

A very important benefit of using in-house personnel for implementing the energy maintenance program is commitment, which is difficult to quantify. Commitment is a factor that can make or break a program, particularly while it is new. Just as senior management must be committed to the program in order for it to succeed, so must the employees who will be performing the maintenance activities. Without commitment, the program will degenerate into just another series of tasks to be performed as quickly as possible before moving on to the next tasks. Commitment is much easier to gain from maintenance employees who are part of the organization and part of the program for the long haul than from contractor employees who may or may not be on this job next year.

Using in-house personnel to implement the program also has its disadvantages. In-house maintenance personnel are the same employees who have been performing or ignoring the maintenance activities in the past. They will have developed a number of habits related to the way in which they perform their duties. Some of these maintenance habits will be bad. If the standard mode of operation has been breakdown maintenance, maintenance personnel will have been conditioned to get in, figure out what is wrong, make the required repairs, and move on to the next maintenance problem. It will be difficult to break these old habits. While there is no guarantee that contract maintenance personnel will be any different, simply by introducing them into a new situation increases the chances that old habits will not be carried over to the new maintenance program.

The energy maintenance program will also require that new activities be performed in the inspection and maintenance of building systems and components. If these activities have been ignored in the past under breakdown maintenance, maintenance personnel may not be properly prepared to perform them efficiently and effectively. Additional training and the development of new job skills will be required, both of which delay program implementation and increase start-up costs. If the program were contracted out, it would be the responsibility of the contractor to provide properly trained maintenance personnel.

In-house personnel may feel that the new program threatens their job security. One of the goals of the program is to reduce overall maintenance costs. To maintenance personnel, reduced costs means reduced labor requirements. Staff will be cut. Overtime will be reduced. The result of these attitudes will be a workforce that not only is not committed to the program, but also will actively work to undermine it. While the maintenance manager can defuse the situation to some extent by actively involving maintenance personnel in the development and implementation of the program, the new program will always be a threat to some. Unfortunately, contracting out program implementation will not solve the problem. In most cases, contracting will only make it worse as now the maintenance workforce will see a clear threat to their job security, particularly if the program succeeds.

When to Use In-House Personnel

There are two conditions that must be met if the energy maintenance program is to be implemented using in-house maintenance personnel: adequate staffing, and the required in-house expertise. Without satisfying both of these conditions, the energy maintenance program will never materialize. Starting the program with too small a staff or one that does not have the proper experience and job skills will result in the program collapsing from within. If the organization cannot commit the personnel to the program, consider contracting all or part of the effort.

How to Implement Energy Maintenance Using In-House Personnel

The first step in implementing an energy maintenance program using in-house personnel is the establishment of the management control structure. Assign the responsibility of managing the program to one person. Without that assigned responsibility, the program will lack the direction and management coordination necessary for it to accomplish its goals and objectives. The program manager must be selected and appointed as early as possible for if that person is to have the responsibility of making the program succeed, he/she must participate in making other implementation decisions.

After the energy maintenance program manager has been appointed, staffing levels must be established. Few maintenance organizations have the luxury of being fully staffed. Budget cutbacks and facility expansions, without corresponding additions to the maintenance staff, have resulted in most maintenance organizations being under-staffed. In spite of this understaffing, maintenance personnel will have to be reassigned from their existing duties to work with the energy maintenance program. How many must be reassigned and what areas they will come from will depend on the goals established for the particular program being implemented.

Rule of Thumb: A commonly used measure when evaluating the effectiveness of maintenance program is the 70–30 rule of thumb. Seventy percent of the work performed by the maintenance staff should be preventive in nature. The remaining 30 percent should be emergency and breakdown maintenance. This same rule can be used as a guide for the energy maintenance program. When the program is fully mature, the organization should adhere to the 70–30 rule. However, when the program is being started, it will be impossible to commit anything near 70 percent of the maintenance resources to energy maintenance due to the number of break-downs that will continue to occur for some time. As the program matures and the number of breakdowns decline, shift additional resources to the energy mainte-nance program.

Rule of Thumb: The recommended staffing level for initiating the energy mainte-nance program is between 15 and 25 percent of the resources available by trade. Too low a level, and the program will never become fully developed. Too high a level, and response time for breakdown maintenance activities will grow unacceptably long.

Do not commit 15 to 25 percent of the entire maintenance department when the program is initiated. Commit personnel in only those maintenance areas where the program is being implemented first. For example, if the first phase of the energy maintenance program includes building chillers and cooling towers, commit only 15 to 25 percent of the time of those personnel who work on that equipment. As the program is expanded to new areas, add maintenance personnel from these areas.

Select personnel based on their ability to contribute to, and their interest in, partici-pating in the energy maintenance program. Meet with employees and explain what the program is and what you want it to accomplish. Review their level of experience and expertise. If the maintenance operation has been one that performs most of its activities in a reactive rather than a preventive mode, it will be necessary to provide additional training to many of these employees. Identify areas where training is needed and schedule those employees for participation.

Once the level of staffing has been determined, the energy maintenance program manager must then develop an estimate of the time required to complete the inspection and maintenance activities for the items being included in the program. Using the tasks identified in earlier chapters, develop a list of equipment and building components that are to be included in the first phase of the program. For each type, estimate the time required to perform the identified tasks. Break the time estimate down by individual trade shops to permit scheduling. Use Figure 17.1 to estimate the number of hours required to complete the identified inspection or maintenance tasks. Use a separate copy of the figure for each type of building system or component being estimated. Use the figure to also record actual times so that estimates may be updated as the energy maintenance program evolves.

Use the estimates of time required to complete the tasks to develop a schedule for the performance of the maintenance tasks. Schedules should be completed weekly based on the projected availability of labor by shop for that week. Do not schedule tasks that exceed 75 percent of the available labor for each week, particularly when the program is in its early stages of implementation. Unforeseen conditions will always lengthen the time required to complete the maintenance tasks. Use Figure 17.2 to develop the weekly schedule of maintenance tasks for the energy maintenance program.

One additional step that is required when implementing an energy maintenance program using in-house personnel is providing the program manager with sufficient administrative support. Maintenance schedules will have to be developed, posted, and tracked. Data will have to be reviewed and filed as it is collected from the field. Work orders will have to be issued and tracked. Building occupants will have to be notified of scheduled equipment outages. Progress and accomplishment reports will have to be written. It is more efficient to use administrative support personnel to complete these required tasks than it is to use maintenance personnel.

HOW TO EFFECTIVELY USE CONTRACTED PROGRAMS

The use of outside contractors to perform maintenance functions has become widely accepted in recent years. Driven by tight budgets and downsizing mandates, a number of maintenance managers have turned to outside firms to perform certain activities that traditionally have been completed using in-house personnel. Activities such as custodial services and landscape maintenance were among the first to be opened up for bidding by outside firms. More recently, contracts have been written for the maintenance of specific and specialized building systems, including chillers, cooling towers, and boilers—some of the key energy using systems that need to be addressed by the energy maintenance program. Contracting these tasks is a cost-effective way of implementing the program quickly and efficiently.

```
┌──────────────────────────────────────────────────────────────────┐
│                          Figure 17.1                               │
│              Energy Maintenance Job Estimate                       │
│                                                                    │
│                                                                    │
│  Building: _____  Room #: _____    │
│                                                                    │
│  Equipment/system description: _____  │
│                                                                    │
│  Type of activity:    ☐ inspection/testing      ☐ maintenance     │
│                  Attach listing of maintenance and inspection      │
│                  activities                                        │
│                                                                    │
│                                                                    │
│                          Estimated Hours    Actual Hours           │
│                                                                    │
│           Electricians:      _____     _____           │
│                                                                    │
│           HVAC mechanics     _____     _____           │
│                                                                    │
│           Plumbers           _____     _____           │
│                                                                    │
│           Painters           _____     _____           │
│                                                                    │
│           Carpenters         _____     _____           │
│                                                                    │
│           Masons             _____     _____           │
│                                                                    │
│           Grounds keepers    _____     _____           │
│                                                                    │
│           _____        _____     _____           │
│                                                                    │
│           _____        _____     _____           │
│                                                                    │
│           _____        _____     _____           │
│                                                                    │
│           _____        _____     _____           │
│                                                                    │
│           _____        _____     _____           │
│                                                                    │
└──────────────────────────────────────────────────────────────────┘
```

Figure 17.2
Maintenance Job Scheduling

Week of: _____

List the hours required by trade shop for each maintenance activity to be performed.

Tasks	Trade Shops			
Total				

Hours available				

The Costs and Benefits of Using a Contracted Program

One of the most important benefits of contracting for the performance of the energy maintenance program is that it establishes a dedicated workforce that is separate from the organization's traditional approach to maintenance. When new programs, such as the energy maintenance program, are implemented that draw labor from existing workforces, it is easy to fall back on the old way of doing things. Pressures in times of crisis to get systems operating will result in managers "temporarily" reassigning maintenance personnel working on the energy maintenance program back to break-down maintenance duties until they catch up on the breakdowns. The problem is that nobody ever catches up. The result is an effective cut in the workforce dedicated to the energy maintenance program.

Contracting for the performance of the maintenance activities associated with the energy maintenance program will all but eliminate this problem. Contract personnel will remain dedicated to the performance of the energy conservation tasks. The existing workforce will remain dedicated to performing breakdown maintenance.

Another benefit of contracting for energy maintenance is flexibility in staffing. Most in-house maintenance forces remain at a fixed level, although work levels vary. Programs that are staffed using in-house maintenance personnel are staffed at a fixed level throughout the year even though the work load may vary. If the in-house maintenance staff is sized for peak levels, the overall costs for the energy maintenance program will be high. If it is sized for average levels, then maintenance will be deferred and some of the benefits of the program will be lost. Contracting for maintenance when implementing the energy maintenance program avoids most of these problems. The contract is written for the performance of specific maintenance tasks. It is the responsibility of the contractor to vary the availability of labor to match the workload.

This flexibility in staffing extends to the tasks performed by the contractor's maintenance personnel as well. For example, many maintenance tasks require that the personnel performing them have special skills. If these skills are required only once or twice a year, the contractor can bring in the necessary personnel when required, perform the maintenance activities, and then move them on to other facilities. If in-house personnel are used, those skills must be taught or learned on the job. If they are only used once or twice a year, the skill level will deteriorate and will never match that of contractor's maintenance personnel who work with those systems on a regular basis.

The most often cited advantage of contracting for maintenance is reduced costs. Proponents claim that lower hourly rates, better skills, and better supervision reduce the overall cost of maintenance in spite of the fact that contractor's costs include a level of profit. The key, they claim, is competition through bidding. Competition forces the contractor to be efficient in both performing and supervising the maintenance activities. It is this lack of competition that has allowed in-house maintenance workforces to become inefficient and expensive.

Contracting for energy maintenance support can reduce total costs as long as the contract can provide what is needed by the organization, when it is needed, and at the required level. Personnel must be properly trained, equipped with the necessary tools, and properly supervised. If the contractor reduces his costs by cutting back on any of these factors, the cost advantages associated with contracting the energy maintenance program will disappear.

Contracting for energy maintenance also has its disadvantages. One of the most serious of these is the focus of the contractor. When the energy maintenance program is managed in-house, decisions are made by considering their long-term consequences. In-house maintenance personnel, as representatives of the owner, are concerned with the impact that their actions will have on both the immediate performance and the long-term life of the component or system. A contractor's focus, however, is somewhat limited by the length of the contract. Profit and terms of the contract are factored into maintenance decisions.

Another disadvantage of contracting for energy maintenance services is it requires significant changes in the way in which the maintenance department operates. You must assume a new role. There are contracts to be written, negotiated, and managed. Complex relationships must be established between a number of parties, including the maintenance manager, in-house maintenance personnel, contract maintenance personnel, and the senior officers of the maintenance contractor. Assuming that maintenance contracts can be managed in the same way that in-house maintenance personnel are managed will only lead to problems in administering the contract.

One of the most serious disadvantages is loss of institutional knowledge. Making the performance of energy maintenance tasks the responsibility of a contractor instead of the in-house maintenance force results in a loss of understanding of how to properly maintain those systems. Changes in contractor personnel, or changes in the contractor itself will result in the loss of this information completely, causing future contractor employees to have to relearn system requirements, at the expense of the maintenance manager.

When to Use Contracts

The decision to use contracts to implement the energy maintenance program is based on a number of factors related to the availability of in-house personnel needed to support the program. If the existing maintenance organization is too small to take on the tasks required to implement the energy maintenance program, contracting for the completion of these tasks is the only option. Similarly, if the skill level of the existing maintenance workforce is not sufficient to complete the tasks required by the energy maintenance program, the program should be contracted out.

Even when the organization has a large enough staff that has the required skills, there are times when it is beneficial to contract out for the implementation of the program. A contract can provide the high level of maintenance staffing support needed to implement the program quickly without having to cut back on breakdown maintenance activities. If speed of implementation is a major consideration, contracts can provide a more rapid response than can be achieved through the use of in-house personnel.

How to Implement Energy Maintenance Using Contracts

Like programs that are run using in-house maintenance personnel, the first step in implementing an energy maintenance program that uses contracts is to establish a management control structure. Begin with the selection of an individual who will have the responsibility of managing the maintenance contract. This person must be familiar with both maintenance and contract management. Select the individual as quickly as possible so that he or she can participate in the development of the maintenance contract.

Using the priorities discussed in Chapter 16, determine the systems and components that are to be contracted for energy maintenance. Which ones will be included in the original phase of the program will depend on the available funding. For bidding purposes, group the items that are to be included in the first phase of the program by the type of personnel required to maintain them. For example, all building chillers can be grouped for bidding as a single contract. If there are a large number of chillers, the group can be further broken down by type or manufacturer. Similarly, building envelope components can be split into several contracts, including one for roofs, one for exterior doors and windows, and one for exterior masonry.

There are two different ways in which the maintenance contracts can be written: fixed price contracts, and time and material contracts. Use fixed price contracts when the scope of the contract and the maintenance tasks are clearly defined. An example of a task for which a fixed price contract would be appropriate is the scraping and painting of all exterior doors and windows in a building. Use time and material contracts where it is not so easy to define the maintenance tasks to be performed, or where performing of the scheduled maintenance activities is likely to identify a number of additional maintenance measures that are in need of completion.

Available funding will establish the level of the maintenance contract or contracts and what tasks are to be performed. As with all energy maintenance programs, start with those items that offer the greatest potential for energy savings. The contract will have to be very specific as to which maintenance tasks are to be performed on what equipment and how frequently. Use the forms from this book as a guide in establishing the list of items to be covered by the contract, what tasks are to be performed, and how often they are to be performed. Identify all data on the systems being maintained that are to be turned over to you by the contractor, and the format that data are to be in.

While the contractors will be supplying the majority of the labor required to perform the maintenance tasks for the energy maintenance program, you, as maintenance manager, will have to establish mechanisms for monitoring the performance of the contractors and controlling the execution of the contracts. Inspectors will have to be assigned to monitor contractor performance on specific tasks, such as the annual chiller inspection, testing, and maintenance. Contractor billings will have to be reviewed and verified before they are paid. Support personnel will be needed to schedule equipment and system outages. Progress and savings estimates will have to be compiled and written.

How to Select the Right Contractor

The process of selecting a contractor is the most critical step in implementing a program for energy maintenance through an outside contractor. Selection of the wrong contractor can prove to be disastrous for the program. The factor that is most often considered when selecting an outside contractor is price. Unfortunately, in many maintenance contractor selection processes, it is the only factor considered. In these cases, the process of procuring maintenance assistance is viewed as being the same as purchasing office supplies: they are all basically the same so cost is the only important factor. The problem is that procuring a maintenance contractor is a far more complicated issue than purchasing items like office supplies. The contractor must be willing to and capable of supporting the energy maintenance effort that is being initiated. In order to determine this, a number of different factors must be examined.

The chances of selecting a contractor that will perform at the level required to support the program can be greatly increased by considering a number of factors during the selection process. Start with contractor qualifications. Does the contractor have the ability to perform as expected under the terms of the energy maintenance contract? The contractor must be able to demonstrate through his or her performance and organization that he/she has the ability to do the work as required. How many maintenance people are employed by the contractor? If the company is too small to support your operation as well as its other clients, the program will suffer. What are the qualifications of the contractor's employees? The employees, like the organization, must be capable of performing the maintenance tasks. What is the turnover rate? High turnover rates are indications of management and organizational problems within the company, problems that may prevent the company from properly performing under the terms of the contract.

Examine the contractor's experience. How long has the contractor been in business? New does not necessarily mean bad, but it is difficult to beat hands-on experience when it comes to maintenance. How many other comparable energy maintenance programs does the contractor currently operate? Nobody wants to be the trainer for a contractor. What type of support structure have they established to manage their field employees and support the contract? Too large a system increases company overhead and tends to crush employees. Too small a system and the contract and employees will not be properly managed. Review their proposed management and support system to determine if it is appropriate.

Finally, have any contractor that is being considered for work on the maintenance program submit a current list of clients, including the name of a client contact. For those contractors that look promising, meet with the client contacts to review contract performance. Meet with maintenance personnel in those organizations to get a better understanding of their perspective on the contractor's performance. Visit the site and examine a number of the systems and components that the contractor is responsible for maintaining. What is their condition? Are they being properly maintained or is the contractor just doing what is required to keep them running?

Develop a list of qualified contractors for bidding on the contract. Prequalify all potential contractors through a two-step bidding process. Require all bidders to submit qualification documentation for review. Client lists and contracts, organizational structure, management schemes, employee credentials—all must be reviewed when qualifying a potential contractor. Only once they are qualified should cost information be reviewed. Use Figure 17.3 to evaluate potential contractors.

One final comment on contracting for the energy maintenance program. Due to the diverse nature of the systems and building components that will be covered by the program, it is best to use multiple contracts with different contractors, each contract designed to focus on a particular set of building systems. Multiple contracts will allow you greater flexibility in meeting the needs of the energy maintenance program.

HOW TO EFFECTIVELY USE HYBRID PROGRAMS

One of the most effective approaches to implementing an energy maintenance program is the hybrid program that makes use of both in-house maintenance personnel and contracts. Program start-up requirements can easily exceed the availability of qualified maintenance personnel. Similarly, the cost of using contracts to implement the energy maintenance program can exceed the budget level for the program. Hybrid programs avoid both problems by balancing the use of in-house personnel with the costs of contracting for the program's implementation.

The Benefits of Using a Hybrid Program

Hybrid energy maintenance programs offer the advantages of both in-house and contracted programs while minimizing the disadvantages of both. Under the hybrid scheme, some of the maintenance tasks required by the energy maintenance program are assigned to in-house maintenance personnel while other tasks, particularly those that the in-house maintenance force is not well equipped to handle, are contracted out. By following this approach, you can take advantage of in-house expertise where it is available, and contract out those areas where there are insufficient in-house resources or expertise.

The hybrid implementation approach is particularly well suited for maintenance managers who in the long term want to perform most or all of the maintenance activities using in-house personnel but lack the personnel to get the program up and running quickly. Contracts can be written to quickly perform the required maintenance on key energy using systems, giving the greatest return possible for the investment in a short period of time. As time and staffing permit, many or all of the maintenance requirements for those systems can be turned over to in-house personnel. The savings produced from the efficiency improvements gained by these systems can be used to fund the next round of contracts that would expand the program to other systems and components.

One of the key advantages of the hybrid implementation scheme is that it allows you to take advantage of the strengths of the in-house maintenance department without overloading it with new maintenance tasks. By reviewing the existing staffing levels

Figure 17.3
Contracter Qualifications

Contractor: _____

Address: _____

_____

_____

Contact person: _____

Phone number: _____

Contractor Ratings

Rate each item on a scale of 1 - 5, with 1 being the lowest

	Rating
Company size	_____
Employee qualifications	_____
Relevant company experience	_____
Operates other energy mainte-nance programs	_____
Management support structure	_____
Client review	_____
_____	_____
_____	_____
_____	_____
_____	_____
Total points	_____

and the projected requirements for the energy management program, you can determine where current resources are sufficient and where they will need to be supplemented by outside forces.

Another advantage offered by the hybrid implementation scheme is that institutional knowledge is not lost as in-house maintenance personnel are kept in close contact with the work the maintenance contractor is performing, particularly if that work is to eventually be taken over by the in-house staff.

How to Implement Energy Maintenance Using Contracts and In-House Personnel

The first step when using this method of implementing an energy maintenance program is the development of the management control structure. Start by assigning the responsibility of managing the program to one person. That manager will oversee both the in-house maintenance personnel and the operation of the maintenance contractor. Again, early appointment will help to get the program started correctly.

Following the energy maintenance program priorities established in earlier chapters, develop a listing of systems and building components that are to be included in the first phase of the program. Review the maintenance tasks that will be performed for each of these items, and determine the skill level and the time required for completion. Review the skills and the number of in-house maintenance personnel who are available to work on the program to determine if the item can be handled in-house or if it should be contracted out.

If in-house maintenance personnel do not have all of the skills required to maintain a particular system, or if there is an insufficient number of in-house personnel available to work on a particular system, consider splitting the maintenance activities required for that system between in-house and contract personnel. For example, one of the most important annual or semi-annual maintenance tasks required for centrifugal chillers is the eddy current testing of the chiller's tubes. Proper performance of the test requires that the operators be skilled in the use of the equipment and the interpretation of the test results. If the organization does not have the equipment or the skills to perform the test, use in-house maintenance personnel to perform the other required maintenance activities and contract personnel to perform the eddy current testing.

As with programs that are operated fully in-house, do not overcommit maintenance personnel to the program. ***Rule of Thumb:*** Limit the in-house participation to no more than 25 percent of the workforce available in any given trade. If additional personnel are required, particularly during program start-up, consider contracting out that portion of the energy maintenance program.

To be most effective, you must develop a balance between in-house staffing and contract support. Contracts should be gradually phased in, starting with the most critical energy using systems. If the long-term goal is to take over most or all of the program so that it will be completed using in-house personnel, make certain that the maintenance force works closely with the contractor's personnel so that they can become familiar with the tasks that must be performed.

How to Select an Implementation Option

Selection of a particular implementation scheme for use in a facility must be based on the program objectives and resources available for that facility. How quickly does the organization expect to see results from the program? What are the goals of the program? What is the relative importance of energy savings and total maintenance cost reduction? What resources are being made available to you for use in starting the program? What are the restrictions on reassigning existing or in hiring additional maintenance personnel? Are there qualified local maintenance contractors who would be able to perform the required activities?

In general, in-house implementation tends to be a slower process than implementation through a maintenance contractor. If speed of implementation is one of the most important considerations in setting up the program, consider the use of maintenance contracts wherever possible to reduce start-up times.

The goals of the program will also help to determine the best implementation scheme. If the major goal is energy conservation, then the quicker the program can be implemented, the quicker the goal can be achieved. Quick implementation calls for the use of a maintenance contract. If the major goal of the program is to reduce total maintenance costs, then implementation of a hybrid program will be the best option.

Programs that are implemented through maintenance contractors generally require more up-front funding than in-house programs. If funding is tight, and start-up funds have not been provided, you will have no choice but to implement the program using in-house maintenance personnel. If some start-up funds are available, consider implementing the program using both in-house and contractor personnel. Hybrid implementations fall between in-house and contractor implementation in both speed of implementation and up-front costs.

SUMMARY

Implementation is a particularly important phase of establishing the maintenance program for energy management. Maintenance managers have several options available. They may implement the program using in-house personnel, an outside contractor, or a combination of both. Which implementation scheme is best suited for a particular application depends on a number of different factors, including the availability of funding, the size of the facility, the speed with which the program must be implemented, and the goals that have been established for the program.

Chapter 18

How to Measure Energy Maintenance Program Effectiveness

Energy management programs that focus on maintenance activities, unlike programs that focus on energy projects, do not end with implementation. They are an ongoing effort that continues to make incremental improvements in energy efficiency long after programs were first put in place. To be most effective, the programs must be dynamic. Facilities and their energy using systems are not static. Conditions change. Priorities change. Any program that is established to manage energy use must be capable of changing along with the facility.

In addition to being able to change, energy management programs must also conduct ongoing self-evaluations. These evaluations must be designed to assess the value of the program and to determine if the program is helping the maintenance operation to achieve the energy conservation and maintenance goals established by the organization.

This chapter focuses on the measures that you can use to develop self-assessment programs. The information presented in this chapter may be used to establish assessment programs in any commercial, industrial, or institutional facility regardless of size or complexity. By following the measures identified, you can develop an assessment program that can be used on an ongoing basis to evaluate energy management and maintenance management effectiveness with respect to the goals established for the program.

Measuring program effectiveness is an ongoing process that is essential to the long-term success of the program. It tells you how effectively the program is operating. It gives hard evidence of the improvements in energy and maintenance efficiency that have resulted from the efforts of the program. It provides a mechanism for tracking actual accomplishments against the goals that were established for the program. It identifies areas where additional resources may be required in order for the program to achieve what was originally intended.

Measuring program effectiveness is also a critical step in maintaining support for the program. If you can demonstrate the program's effectiveness through hard data, it decreases the chances that funding for the effort will be reduced. The data developed by measuring effectiveness can even be used to justify increases in funding to accelerate program implementation or to expand program operations.

There are a number of different measures of effectiveness that may be used in assessing maintenance programs for energy conservation. Selection of particular measures will depend on the primary goals of the maintenance program and the type of organization where it is being applied. Some measures are better suited for industrial applications, while others provide more usable information in commercial or institutional facilities.

There are three different ways in which the effectiveness of the maintenance program can be measured and monitored over time: energy accounting, maintenance tracking, and benchmarking. All three evaluation methods may be used to identify how well the program is currently performing and what remains to be accomplished. The data produced may be used to quantify savings that the program has produced, a requirement if you are to retain the support of senior management for the program.

ENERGY ACCOUNTING

The primary goal of the maintenance program is to reduce the energy requirements of a facility. In order to quantify the energy savings that have resulted from the implementation of the program, you must establish an energy accounting system that tracks the energy performance of the facility and provides a mechanism for identifying reductions that are a direct result of the program. The problem facing you in setting up the accounting system is that there are many variables in the operation of a facility that will impact energy use, including weather, use, occupancy, and production. The system, if it is to be an effective indicator of the performance of the energy program, must account for the variables that impact energy use in that particular facility. By doing so, any remaining variations may be attributed to the maintenance program.

All effective energy performance measures deal with energy use rather than costs. Costs simply vary too much as the result of price fluctuations in energy pricing to serve as an effective performance measure. Therefore, energy performance measures are nearly always presented in terms of the Btu energy use of the facility. Some are simple to calculate, while others are complex, requiring statistical modeling. Not all can be applied to all types of facilities. The key is to select the one that is most appropriate for use in your particular facility.

Use Figure 18.1 to determine the total monthly and annual energy use of a facility in thousands of Btu. The figure lists average conversion factors for various fuels that may be used to determine their Btu equivalents. If the actual Btu conversion factor is known for fuels such as natural gas or coal, use that factor instead. Information compiled using the figure will be used in determining various energy performance measures.

Figure 18.1
Facility Energy Use Log

Building: _____   Year: _____

Fuel	Jan	Feb	Mar	Apr	May	June	July	Aug	Sept	Oct	Nov	Dec	Total
Electricity: kWh													
10^3 Btu													
Natural Gas: cf													
10^3 Btu													
Propane: lbs													
10^3 Btu													
#2 Oil: gal													
10^3 Btu													
#6 Oil: gal													
10^3 Btu													
Coal: tons													
10^3 Btu													
Total Btu (10^3)													

Btu Conversion Factors

Electricity:	multiply kWh by 3.413	#2 Oil:	multiply gal by 139
Natural Gas:	multiply cf by 1.0	#6 Oil:	multiply gal by 151
Propane:	multiply lbs by 21.5	Coal:	multiply tons by 12

For all energy efficiency calculations, collect data for a base year before any of the maintenance measures have been implemented, preferably for the year preceding the implementation of the maintenance program. Complete a copy of Figure 18.1 using these data and mark it "Base Year." These data will be used in evaluating program effectiveness using any of the energy performance indicators. Comparisons between data collected after implementation and the data from the base year will produce estimates of the energy savings that have been achieved as a result of the maintenance program.

Energy Use per Square Foot

One of the simplest to calculate and most widely used energy performance measures is the total energy use of the facility in Btu divided by the facility's total occupied area. Expressed in Btu per square foot, this energy performance indicator is easy to calculate and use. By normalizing the energy use by the occupied area of the facility, it provides a mechanism for correcting energy use figures for changes in the size of the facility over time.

There are two significant drawbacks to the use of the Btu per square foot performance measure. First, it assumes that the most important factor in determining energy use is the occupied area of the facility. The impact of other factors, such as weather, occupancy, and process energy requirements, is assumed to be low. However, in practice, energy use in nearly all facilities is influenced by a wide range of factors. Assuming that area is the only important one limits the effectiveness of the performance measure.

The second significant drawback in the use of the Btu per square foot performance indicator is its assumption that energy use is linearly proportional to the area of the facility. If the area of the facility increases by 10 percent, energy use, it is assumed, will also increase by 10 percent. In practice, energy use does not vary linearly with area. A doubling of area typically results in an increase in energy use of between one-half and two-thirds, depending on the type of construction and the use of the facility.

In spite of these drawbacks, the Btu per square foot performance measure can be used successfully in a number of different applications. Three conditions must be met. First, facility process energy requirements must be small, typically no more than 10 percent of the facility's total energy use. Second, weather factors must be relatively constant from year to year, also varying by no more than 10 percent. Third, the area of the facility must not change considerably during the time when energy performance is being evaluated. If these three conditions can be met, the Btu per square foot performance measure will provide a good and easy-to-calculate indication of the energy efficiency improvement achieved through the implementation of sound maintenance practices.

To determine and track the energy savings produced by the maintenance program on a Btu per square foot basis, use two copies of Figure 18.2. Complete one copy of the figure using the base year data recorded on Figure 18.1. Do not complete the calculation at the bottom of the figure. Complete a second copy of Figure 18.2 using the data from the current year. To determine the percentage of energy savings on a Btu per square foot basis, enter the annual total Btu per square foot for the current and base years in the appropriate boxes at the bottom of the figure. Subtract the base year value from the current year value, and divide by the base year value to determine the percentage of savings.

Energy Use per Unit of Production

In manufacturing and industrial facilities, where production of a product is the primary activity carried out, energy efficiency can be measured by relating energy use to production output of the facility. For example, an automobile assembly plant could measure the energy efficiency of the facility by calculating the Btu energy use per vehicle produced. Similarly, a food processing plant's measure of energy efficiency could be its Btu energy use per ton of food processed.

The Btu per unit of production may also be applied effectively to other facilities that are not normally considered to be production or manufacturing facilities, such as a school or a university. Energy efficiency in these facilities is determined by dividing the energy use in Btu by the number of students attending. Hospitals could use Btu per occupied bed as an efficiency measure. Outpatient facilities could determine their Btu per procedure. Similar efficiency measures can be calculated for other facilities depending on the processes that take place there.

The Btu per unit of production has the advantage of rating energy efficiency in terms of what is most important to the operation of the facility—its output. It is easy to use and provides a good basis for comparison of the energy use of the facility over time. It does, however, have its limitations. The most serious of these is the assumption that energy use is directly proportional to production output of the facility. For energy use to be proportional to production output, the energy use resulting from other factors, such as weather, must be relatively low. The energy use must also increase linearly with production.

A second limitation of the Btu per unit of production efficiency measure is the difficulty encountered when defining what a unit of production is. In some industrial and manufacturing facilities, the product produced remains constant: steel rods, computer chips, paper, and so forth. In many others, it varies. With that variation in produced products comes a variation in energy requirements, a variation that will make it difficult to determine how much energy is being saved as a result of the energy management program.

If, however, the product produced by the facility remains fairly constant, and the majority of the energy used by the facility is for process requirements, the Btu per unit of production measure is a very good indicator of energy efficiency.

To determine and track the energy savings produced by the maintenance program on a Btu per unit of production basis, use two copies of Figure 18.3. Complete one copy of the figure using the base year data recorded on Figure 18.1. Do not complete

Figure 18.2
Facility Btu per Square Foot Worksheet

Building: _____   Year: _____

	Energy Use $(10^3$ Btu)	Area (sq ft)	10^3 Btu/sq ft
January			
February			
March			
April			
May			
June			
July			
August			
September			
October			
November			
December			
Annual total			

Percent Reduction in Energy Use

$$\frac{\boxed{} - \boxed{}}{\boxed{}} = \boxed{}$$

current year total (Btu/sq-ft)   base year total (Btu/sq-ft)   % reduction

base year total (Btu/sq-ft)

the calculation at the bottom of the figure. Complete a second copy of Figure 18.3 using the data from the current year. To determine the percentage of energy savings on a Btu per unit of production basis, enter the total Btu per unit of production for the current and base years in the appropriate boxes at the bottom of the figure. Subtract the base year value from the current year value, and divide by the base year value to determine the percentage of savings.

Energy Use per Heating Degree Day

An effective way to measure the impact that weather has on a facility's heating energy use is the heating degree day (HDD). The HDD is defined as the number of degrees that the daily mean temperature is below 65 degrees Fahrenheit. For example, if the mean temperature for a particular day is 40 degrees Fahrenheit, the number of heating degree days for that day is 25. By adding the number of HDDs in a month and a year and comparing it to the average value for that month or for the average year, the relative severity of that month or year can be estimated.

In facilities where building heating loads represent the major component of energy use, the energy efficiency of the facility can be determined by relating the energy use to the total number of HDD in a particular month or for the year. Expressed in Btu per HDD, this efficiency measure is easy to calculate. Since the energy use is normalized by a measure of the severity of the heating season, it provides a mechanism for comparing energy use during different months and years.

The Btu per HDD efficiency measure is accurate only when the majority of the energy use in the facility is for heating purposes. Examples of facilities where heating energy requirements are the major energy load include warehouses and non–air conditioned, light industrial facilities. The measure loses accuracy as other energy loads, including lighting, cooling, and production requirements, become a significant portion of the total energy use of the facility.

To determine and track the energy savings produced by the maintenance program based on Btu per heating degree day, use two copies of Figure 18.4. Complete one copy of the figure using the base year data recorded on Figure 18.1. Do not complete the calculation at the bottom of the figure. Complete a second copy of Figure 18.4 using the data from the current year. To determine the percentage of energy savings using Btu per heating degree day, enter the total Btu per heating degree day for the current and base years in the appropriate boxes at the bottom of the figure. Subtract the base year value from the current year value, and divide by the base year value to determine the percentage of savings.

Statistical Models of Energy Use

Each of the previous estimates of energy efficiency can be used effectively in facilities that meet certain conditions. Each of these estimates assumes that the majority of energy use in a particular facility can be attributed to a single load variable, such as building

Figure 18.3
Facility Btu per Unit of Production Worksheet

Building: _____ Year: _____

	Energy Use (10^3 Btu)	Number of Units Produced	10^3 Btu/sq ft
January			
February			
March			
April			
May			
June			
July			
August			
September			
October			
November			
December			
Annual total			

Percent Reduction in Energy Use

$$\frac{\boxed{} \text{current year (Btu/unit)} - \boxed{} \text{base year (Btu/unit)}}{\boxed{} \text{base year (Btu/unit)}} = \boxed{} \text{\% reduction}$$

Figure 18.4
Facility Btu per Heating Degree Day Worksheet

Building: _____ Year: _____

	Energy Use (10^3 Btu)	Heating Degree Days	10^3 Btu/HDD
January			
February			
March			
April			
May			
June			
July			
August			
September			
October			
November			
December			
Annual total			

Percent Reduction in Energy Use

$$\frac{\text{current year (Btu/HDD)} \quad - \quad \text{base year (Btu/HDD)}}{\text{base year (Btu/HDD)}} \quad = \quad \boxed{} \text{ \% reduction}$$

area, units of production, or weather factors. In most facilities, however, energy use is determined not by a single factor, but rather by multiple factors. The area of the facility, the severity of the weather, the number of days that the facility is open, the number of units produced—all contribute to the total energy use of the facility. In order to make the estimate of energy efficiency improvement resulting from the energy management program as accurate as possible, all of these factors must be taken into consideration by the energy accounting system.

One method that allows users to determine the impact of multiple factors on energy use is a statistical model called multiple linear regression. Multiple linear regression uses historical data to determine the relationship between energy use and any number of different factors. Not only does it determine those relationships, it also identifies what fraction of total energy use is the result of that particular factor. Although any number of factors can be used to analyze energy use, in most applications, the number of factors is generally limited to three or four that are considered to be the most important for that particular facility.

For example, the energy use in a typical office building is a function of the area of the building, how many days the building is open, the severity of the heating season, and the severity of the cooling season. The multiple linear regression model would express this energy use as the following:

$$\text{Use} = a + b \times (\text{area}) + c \times (\text{days open}) + d \times (\text{heating degree days}) + e \times c\ (\text{cooling degree days})$$

where
    area = the area of the facility
    days open = the number of days that month that the facility was open
    heating degree days = the number of heating degree days that month
    cooling degree days = the number of cooling degree days that month

By entering two or more years' worth of monthly data for energy use, days open, heating degree days, and cooling degree days, the regression model would determine the values for the constants a, b, c, d, and e. The energy savings produced by the maintenance program could then be determined by comparing the actual energy use of the facility to what was predicted by the regression model.

Regression models are extremely complex in how they operate, and are beyond the scope of this text. Fortunately, a number of low-cost computer-based regression programs are available that can be used effectively to perform the energy analysis required in order to determine the savings being produced by the maintenance program. All you must do is select the particular model to use, and enter the data necessary in order to calculate the regression's constants. Maintenance managers are encouraged to review several of these programs and select one that best suits their needs.

The use of a statistical model in evaluating the energy efficiency of a maintenance program can provide very accurate data. Although the analysis is more complex, the use of a standard computer program simplifies the process.

To determine and track the energy savings produced by the maintenance program, several years of energy use data from before the implementation of the program will have to be collected and entered into the regression program. At a minimum, two years of monthly data will be required. If data are available for three years, this will help refine the regression estimate. Use several copies of Figure 18.1 to record and calculate the data. Additional data, including facility area, occupancy information, heating degree days, cooling degree days, and so forth, will have to be collected for each month.

Enter the monthly energy use totals and the other monthly data into the regression program. The program will determine the equation that relates energy use in the facility to the various factors, such as area and days the facility is occupied. As additional data are collected once the maintenance program has been implemented, enter those values into the regression equation to estimate what the energy use would have been if the program had not been initiated. The difference between the estimate and the actual energy use is the energy savings that are a result of the maintenance program.

One final comment concerning the regression model. In addition to showing how energy use varies with changes in a number of different factors, the model can be used to determine a very important energy conservation value, the base energy use of the facility. The base energy use is important because it represents the energy that the facility would use if all other factors were eliminated. In most facilities, base energy use represents between one-third and one-half of the facility's total energy use.

In the example cited above, energy use is determined from the equation:

$$\text{Use} = a + b \times (\text{area}) + c \times (\text{days open}) + d \times (\text{heating degree days}) + e \times (\text{cooling degree days})$$

If the number of days that the facility was open, the number of heating degree days, and the number of cooling degree days were all zero, the equation would become:

$$\text{Use} = a + B \times (\text{area})$$

This is the base energy use of the facility. It is often in this base energy use where the potential for energy savings is high. If you use the statistical model, you must pay particular attention to base energy use and watch how it varies with the implementation of maintenance and energy conservation activities.

MAINTENANCE PERFORMANCE TRACKING

Although energy management is a primary goal of the maintenance program, it is only one of several goals. Other goals for the program include reduction in equipment downtime, improvement of system reliability, reduction in emergency maintenance requirements, and reduction in total maintenance costs. To determine if the maintenance program is effectively meeting these goals, you must track the performance of the program in these areas.

Unlike energy accounting systems, where the Btu equivalents and other factors are easily defined and universally accepted, maintenance tracking systems use terms that mean different things to different people. How one person interprets the meaning of a particular term can significantly alter how the maintenance tracking system operates. For example, in some maintenance programs, all maintenance activities, including those that are preventive, are handled through work orders. In other programs, preventive and scheduled maintenance activities are handled through a system that is separate from the work order system. Tracking performance by counting the number of work orders completed in a given month would give entirely different results in the two systems. While it is not necessary for all maintenance organizations to adopt the same system and definitions, it is important that everyone within a particular organization understand what is being tracked and evaluated.

All maintenance tracking will require that a review be made of historical maintenance data in order to establish a performance baseline. All baseline data must be from the time period before the maintenance program for energy management was established. It will be used to establish a level of historical maintenance performance. Deviations from this historical base after the maintenance program has been implemented can be attributed to improvements made by the program, assuming all other major factors remain the same.

Another major difference between energy accounting systems and maintenance performance tracking systems is the minimum period of time that may be used to compare performance. For energy accounting systems, the minimum time period is typically one month; energy use in a given month can be compared to energy use in the same month of another year. Maintenance performance tracking systems cannot make comparisons based on such a short time period as the breakdown of a single piece of equipment can artificially skew the data. Therefore, it is recommended that comparisons of the results of the performance tracking system be kept to a minimum of every six months. More accurate results will be obtained if comparisons are made on a yearly basis.

There are a number of different factors that can be used to measure maintenance performance. Most deal with frequency of events, such as equipment breakdowns, or number of labor hours involved in breakdown maintenance. It is recommended that comparisons not be made on the basis of cost since variations in labor rates will alter the results of the analysis. The cost avoidance resulting from the implementation of the maintenance program can be determined once the reduction in breakdown frequency, lost production time, and time spent on breakdown maintenance have been determined.

Breakdown Frequency

The cost of equipment downtime is often underestimated as maintenance managers factor in only the cost of the labor and materials required to get the equipment back operating. What is usually overlooked are the indirect costs resulting from the equipment's failure: lost manufacturing and production capabilities, displaced personnel and operations, decreased worker productivity, and canceled operations. In industrial and manufacturing facilities, equipment downtime is the single biggest maintenance cost. In commercial and institutional facilities, it is more difficult to quantify, but just as significant. For these reasons, one of the most effective measures of maintenance performance is the frequency with which equipment breaks down. Tracking the frequency of breakdowns can be used effectively in all types of facilities.

Before setting up a system to track breakdowns, it is essential that what a constitutes a breakdown is clearly defined and understood by everyone. If different definitions are used of what a breakdown is, the results of the maintenance tracking program will be unreliable. Therefore, you must develop first a working definition of what a breakdown is and put in place the mechanism to apply it consistently in their tracking program. While the definition will vary from organization to organization, it will have several common characteristics. Breakdowns are equipment or system failures that result in the interruption of services to all or part of the facility. A minimum period of time, such as one hour, may be specified for the interruption of services to qualify as a breakdown. Correction of the breakdown must require the use of maintenance personnel. Equipment stoppages resulting from power failures or momentary power outages, such as those caused by storms, are not considered to be equipment breakdowns.

Tracking breakdowns can provide more useful information if it is broken down by a number of categories, such as mechanical equipment, electrical equipment, the building envelope, and the building interior. Further separating these categories into different groups, particularly while the program is still being developed, will provide an even greater detail of information. For example, mechanical equipment breakdowns can be tracked by separate logs for chillers, cooling towers, boilers, air handlers, and so forth.

In order to establish a baseline for breakdown frequencies, maintenance records will have to be reviewed for a one-to-three-year period. Using the definition of what a breakdown is, record their frequency on Figure 18.5 in the column marked Base Year. Use a separate figure for each type of system or component being logged. The figure has three additional columns that may be used to record breakdown frequencies for three separate years after the maintenance program has been implemented. Enter the year of the data at the top of the column. The last line of each column is used to calculate the percentage change in the number of breakdowns from the base year.

Maintenance Requests

Another relatively simple method for evaluating maintenance performance is by tracking the number of maintenance requests that are received by the maintenance department from building occupants. One of the purposes of the maintenance program is to reduce the number of system and component failures through the performance of scheduled and preventive maintenance. If scheduled and preventive maintenance are functioning effectively, problems will be avoided or repaired before building occupants become aware of them. Therefore, the number of maintenance requests reported by building occupants can be used to gauge the effectiveness of the maintenance program.

Figure 18.5
Maintenance Breakdown Log

Building: _____

Type of system/component: _____

	Frequency of Breakdowns			
	Base Year			
January				
February				
March				
April				
May				
June				
July				
August				
September				
October				
November				
December				
Total				
% Change				

Tracking the number of maintenance requests is most effective in commercial and institutional facilities.

Successful tracking of the number of maintenance requests requires a number of things. The maintenance department must have a formal mechanism for building occupants to report maintenance problems, and the building occupants must be willing to use it. In facilities where problems are often verbally reported to maintenance personnel as they make their rounds through the facility, and where no written record is made, the system cannot effectively track their frequency.

The system used for processing maintenance requests must have a mechanism for differentiating between those made by building occupants and maintenance personnel. Similarly, the system must differentiate between scheduled and preventive maintenance activities and those that are generated by building occupants.

If the maintenance request system can meet these conditions, tracking the number of maintenance requests originated by the building occupants can be an effective measure of maintenance system performance. As with other systems designed to evaluate maintenance performance, dividing the requests into a number of different categories based on what was the cause for the maintenance action will help to identify areas where performance is improving, and areas where additional efforts must be made. Typical categories include mechanical systems, electrical equipment, architectural finishes, and structural systems.

Start with a review of maintenance records for a one-to-three-year period before the maintenance program was implemented, identifying the number of maintenance requests that were received from building occupants. Record the number in the column marked Base Year in Figure 18.6. Use a separate copy of the figure for each category being logged. The figure has three additional columns that may be used to record the number of maintenance requests for three separate years after the maintenance program has been implemented. Enter the year of the data at the top of the column. The last line of each column is used to calculate the percentage change in the number of maintenance requests received from the base year.

Breakdown Maintenance Overtime Hours

One of the most significant expenses of performing breakdown maintenance is the labor cost for overtime. While not all breakdown maintenance is performed on overtime, and not all overtime worked in an organization is the result of performing breakdown maintenance, the number of overtime hours spent performing breakdown maintenance can be used as an indicator of the efficiency of the maintenance program. Emergency maintenance overtime tracking systems work equally well in all types of facilities.

As with systems that track the frequency of breakdowns, emergency maintenance overtime tracking systems must start with a clear, concise definition of what a breakdown is. That definition must then be applied consistently when determining what overtime hours to track by the performance indicator. For example, overtime hours spent performing routine or preventive maintenance activities would not be counted

Figure 18.6
Maintenance Requests from Occupants

Building: _____

Type of system/component: _____

	Number of Maintenance Requests			
	Base Year			
January				
February				
March				
April				
May				
June				
July				
August				
September				
October				
November				
December				
Total				
% Change				

by the system. Similarly, overtime hours spent restarting mechanical systems after power outages caused by storms or utility company failures would also not be counted. However, overtime hours required as the result power outages caused by the failure of electrical equipment owned and maintained by the facility would be added to the hours tracked by the overtime performance indicator.

Tracking breakdown maintenance overtime by a number of different categories will help to identify areas in which large numbers of hours of overtime are currently being used. Separate the overtime hours into several categories based on what was the cause of the need to use overtime, such as failures mechanical systems, electrical systems, and building structural systems. Log all overtime required as a result of that particular breakdown against that category regardless of what maintenance trades were required to perform the overtime. For example, if a plumbing system failure caused a flood in an area of a building, and both plumbers and custodial personnel were required to perform overtime work as a result, the total of the overtime hours for both groups would be logged against the mechanical system category.

As with other tracking systems, a baseline must be established for overtime hours. Review maintenance overtime records for a one-to-three-year period. Using the definition of what a breakdown is, and what overtime hours are to be included in the tracking program, record their frequency on Figure 18.7 in the column marked Base Year. Use a separate figure for each category being logged. The figure has three additional columns that may be used to record breakdown overtime for three separate years after the maintenance program has been implemented. Enter the year of the data at the top of the column. The last line of each column may be used to calculate the percentage change in the number of breakdown overtime hours required from the base year.

Availability Factor

A measure of how effective a maintenance program is at keeping equipment and systems operating is its availability factor. The availability factor for a particular system or piece of equipment is the ratio of the number of hours that it was operating or available to operate to the total number of hours in the reporting period. For example, if a building chiller was out of service for 40 hours during a cooling season that ran 1,500 hours long, its availability factor would be 1,460 hours divided by 1,500 hours, or 0.9733.

Although availability factors have been most widely applied in industrial and manufacturing facilities where the availability of equipment is a critical factor in keeping production lines operating, the availability factor can be effectively applied to critical systems in all facilities. For example, you can use an availability factor to measure the effectiveness of their maintenance program on systems such as building chillers, boilers, and essential air handlers.

One significant difference between the availability factor and other maintenance performance measures are their scope. Availability factors are applied at the individual system and component level while all other maintenance performance measures are applied across categories of systems or all systems. Availability factors serve as a

Figure 18.7
Breakdown Maintenance Overtime Log

Building: _____

Type of system/component: _____

	Breakdown Maintenance Overtime			
	Base Year			
January				
February				
March				
April				
May				
June				
July				
August				
September				
October				
November				
December				
Total				
% Change				

measure of the performance of the maintenance program with respect to that particular item. Other maintenance performance indicators measure the overall performance of the entire maintenance program.

For systems and components that are required year round, availability factors are most effective when they are applied on an annual basis. For systems that are required on a monthly or seasonal basis, they are best applied on a monthly basis. As with other maintenance performance tracking systems, a baseline must be established. Review maintenance records for a one-to-three-year period prior to implementation of the maintenance program, and determine the availability factor for the particular system or component being tracked.

Use Figure 18.8 to determine the availability factor for particular equipment and systems. For the year prior to implementation of the maintenance program, enter the number of hours that the system or component was out of service for each month. Enter the total number of hours per month that the system or component should have been available for use. Total both the hours out of service and the total hours. Enter the total values in the appropriate boxes at the bottom of the figure to determine the availability factor for the system or component. After the maintenance program for energy management has been implemented on that particular item, complete a new copy of the figure and calculate the new availability factor. Improvements in availability factors can be attributed to the maintenance program.

BENCHMARKING ENERGY AND MAINTENANCE PERFORMANCE DATA

Energy and maintenance performance factors are useful in determining how effective the maintenance program has been in reducing energy and maintenance requirements in the facility with respect to its historical performance. While they can tell if the maintenance operation is becoming more or less efficient, they cannot tell how much room there is for improvement in performance. Energy and maintenance performance factors are all relative to the existing operation. What is needed is a means of measuring the efficiency of their operation relative to other similar facilities—ideally the best and most efficient comparable facilities. By making this comparison, you can effectively determine not only how efficient their operation is but also how much room there is for improvement.

One technique that can be used to develop this information is benchmarking. Benchmarking is the process of looking for and identifying the methods and processes that can be adapted by an organization to improve its performance. It has traditionally been used by managers to compare their organization's performance with those who are recognized as being world-class market leaders, in an attempt to find ways to improve performance. For energy management and maintenance operations, benchmarking can be used to evaluate the energy and maintenance efficiency of the organization with respect to other organizations, including those who are considered to be world-class energy and maintenance organizations.

Figure 18.8
Availability Factor Worksheet

Building: _____

Component/system: _____   Year: _____

	Hours Out of Service	Total Hours
January		
February		
March		
April		
May		
June		
July		
August		
September		
October		
November		
December		
Total		

$\boxed{}$ - $\boxed{}$

Annual total hours   Annual hours out of service

$= \boxed{}$

$\boxed{}$ Availability factor

Annual total hours

There are four major outcomes of benchmarking energy management and maintenance operations. Benchmarking will identify efficient practices and processes in other organizations. It will show when additional improvements can be made. It will help managers establish attainable performance goals to improve their operations. It will identify specific activities that can be used to improve the performance of those energy and maintenance operations. For the energy and maintenance managers looking to evaluate their operations, the most important of these are identifying when additional efficiency improvements can be made, and quantifying what those improvements can be through the establishment of new goals.

The performance indicators used in gathering benchmarking data are typically the same ones used to gauge energy and maintenance efficiency within the organization. Energy use per square foot, energy use per unit of production, energy use per heating degree day, breakdown frequencies, the number of maintenance requests, breakdown maintenance overtime requirements, and equipment availability factors all can work effectively as benchmarking energy and maintenance performance data. Which performance indicator is best for a particular benchmarking process depends on the approach taken to benchmarking.

Approaches to Benchmarking

There are a several different benchmarking approaches that you can take to identify areas where improvements can be made in your operating efficiency. While all benchmarking approaches can result in improvements being made, some are better suited for certain types of organizations. In most cases, it is best to mix and match the approaches based on the needs of the organization and the availability of benchmarking data.

One benchmarking approach that is very effective in examining energy and maintenance efficiency in facilities with a number of different buildings at the same location or at different locations, is the internal approach. If individual buildings are metered for energy use, any of the energy accounting procedures identified earlier in this chapter may be used to analyze energy use within the buildings and to identify those buildings with the highest energy efficiency. Similarly, if one system tracks all maintenance activities in those buildings, any of the maintenance performance tracking procedures identified earlier in this chapter may be used to analyze the effectiveness of the maintenance practices in those buildings and to identify the ones with the better maintenance practices.

The goal of internal benchmarking is not to find the best building, but rather to find out if there are some differences in operating practices between buildings that make some of them more efficient than others. Equally important, internal benchmarking strives to determine whether the differences in operating practices in those more efficient buildings can be adapted to work in less efficient buildings, thus raising the overall efficiency of the facility.

Even for facilities that plan to look to outside organizations for benchmarking data, internal benchmarking is a necessary starting point. It provides you with the needed performance data on your organization that will be used to compare your operation to that of other facilities.

Start by using one or more of the energy accounting and maintenance performance measures identified earlier in this chapter. **_Rule of Thumb:_** If the number of buildings involved is ten or less, use the most appropriate measures to determine the range of efficiencies for the buildings. If there are more than ten buildings, a representative sample of the buildings may have to be used, although it is still best to evaluate the energy and maintenance efficiency of all buildings individually. Rank the buildings in order of their performance, then study the operation of the least and most efficient buildings. What is different about them? Accounting for differences in age, function, construction, and so on, what makes one building more efficient than the other? What are the differences in operating and maintenance practices between the buildings? Can the practices employed in the more efficient buildings be adapted for use in the less efficient ones? Internal benchmarking can provide valuable insights into how to improve overall operating efficiency for the facility.

Internal benchmarking for energy efficiency and maintenance effectiveness can help you identify practices already employed in some buildings to improve operating efficiency in others, but it cannot tell you how efficient the operations are with respect to other facilities. The energy and maintenance managers who only look at energy efficiency and maintenance practices within their own organizations will have no idea if their operation is as efficient as it could be or if improvements could be made. Industrywide benchmarking will provide you with the necessary information to evaluate your organization's energy efficiency and maintenance practices with respect to how well others in the industry are performing.

Industrywide benchmarking examines the operation of a wide range of related organizations to identify those that are operating most efficiently. The organizations that are selected for benchmarking typically produce similar products or services. Use the same performance indicators used in determining the energy and maintenance efficiency within your own organization, and compare the results. When applying industrywide benchmarking, managers must realize that significant differences in how the facilities are operated, the climates where they are located, and the products or services they produce will result in widely varying measures of efficiency.

Steps in Benchmarking

Benchmarking energy management and maintenance operations is more than simply gathering energy use and maintenance operations data from other organizations and comparing them to data from your organization. Successful benchmarking requires hard work, starting with research. One of the most difficult steps is the identification of similar organizations. On the surface, many organizations look identical, but when the details of their operations are examined, they turn out to be very different.

Start by identifying organizations that are in the same business as yours. Meet with their energy and maintenance managers. Walk the facility. Does the facility support the same type of operations that are supported in your facility? Does it have the same operating hours? What differences are there in weather-related energy use factors between the facilities? Are their energy using systems similar? Look at their energy management and maintenance processes. Have they implemented programs to manage both? What is the status of those programs with respect to the programs in your facility?

Once similar facilities have been identified, data will have to be collected on the operation of those organizations, including the energy accounting and maintenance efficiency measures identified earlier in this chapter. Study the various performance measures and identify areas where gaps in performance exist. For example, if two office buildings have been identified as being similar in operation, size, and construction, and weather factors do not vary significantly between the two facilities, the two facilities should have similar Btu per square foot energy use values. If there are differences, the other facility will have to be closely examined to determine the factors that result in the difference in performance. If those factors can be applied to your facility, then new target values can be established for your energy management program.

Benchmarking is not an easy task that can be entered into lightly. It is hard work. It requires the development of an understanding of what makes practices efficient and it requires that you develop an understanding of which of those practices can be best adopted for your organization. It also does not end with the identification of more efficient processes or the development of new efficiency targets. Benchmarking in an ongoing process that must continually reassess the energy and maintenance efficiency of an operation. Operations are assessed and benchmarked against others. New processes are adopted, and the results must once again be assessed and benchmarked against others. Remember, facilities are always changing. If their operations are to remain efficient, energy and maintenance managers must always be looking for ways to improve. Benchmarking is one way to assess the operation and to identify ways in which improvements can be made through change.

SUMMARY

Energy management efforts do not end with the implementation of the program. Energy management is a process that requires constant effort and attention. Part of that effort must be directed to monitoring and evaluating the performance of the program. Without feedback, it is impossible to develop programs that will achieve their maximum potential savings.

The effectiveness of energy and maintenance management programs is best monitored and evaluated through the use of energy accounting and maintenance tracking systems. Measures of operating efficiency are then calculated for the facility for comparison with historical data as well as with data from other, comparable facilities. Benchmarking practices can be used to evaluate the performance of the programs with respect to programs in place in other facilities and to identify additional activities that can be employed to further improve operating efficiency.

The results of the energy accounting, maintenance tracking, and benchmarking efforts can then be used to establish new goals for the energy and maintenance program, and to track the program's effectiveness in achieving those goals.

Appendix

Energy Conversion of Energy Terms
Glossary of Energy Terms

GLOSSARY OF ENERGY TERMS ————————————————————

Absorption chiller. A refrigeration unit that uses a source of heat to power the refrigeration cycle. Typical heat sources include steam and hot water.

Active power. The component of power in an alternating current circuit that is converted into useful work. The ratio of the active power in a circuit to its apparent power is the power factor. *See also* Power factor.

Average rated life (lamps). The number of hours that a large group of lamps can be operated before one half of them fail. While the actual life varies between lamps of the same type and manufacturer, the average rated lamp life is used as a measure of the life expectancy of a particular lamp type.

Ballast. A current limiting device used in the operation of electric discharge lamps, including fluorescent, mercury vapor, metal-halide, and high pressure sodium. Ballasts may be magnetic or electronic in design.

Base load. The energy requirements of a facility that are independent of weather, occupancy, production, and other variations in operation.

Blowdown. Water that is bled from boiler and cooling tower water systems to reduce the level of total dissolved solids.

Boiler, high pressure. A boiler connected to a steam system that operates at pressures greater than 15 psi or one that is connected to a hot water system that operates at pressures greater than 160 psi or temperatures greater than 250° F.

Boiler horsepower. A capacity rating for boilers. One horsepower equals 33,475 Btu per hour.

Boiler, low pressure. A boiler that is connected to a steam system that has a maximum

Figure A.1
Energy Conversion Factors

To Convert From	Multiply By	To Obtain
Barrels of oil	42	Gallons
Btu	0.000293	Kilowatt hours
Btu/hour	0.2928	Watts
CCF natural gas	103,000	Btu
Gallon #2 oil	138,690	Btu
Gallon #5 oil	6,287,000	Btu
Gallon #6 oil	6,286,980	Btu
Horsepower (boiler)	33,475	Btu/hour
MCF natural gas	1,000	Cubic feet
Kilowatt hours	3,413	Btu
Kilowatts	56.92	Btu/minute
Short ton coal	24,500,000	Btu
Therms natural gas	100,000	Btu
Watts	3.413	Btu/hour

operating pressure of 15 psi or a hot water system that has a maximum operating pressure of 160 psi and a maximum operating temperature of 250° F.

British thermal unit (BTU). The heat required to raise the temperature of one pound of water one degree Fahrenheit.

Building envelope. All external surfaces of a building, including the roof, exterior walls, doors, windows, below grade walls, and floors.

Centrifugal chiller. A refrigeration unit that uses mechanical energy to drive a centrifugal compressor to generate chilled water. Typical drives include electric motors and steam turbines.

Chilled water. The water that is circulated between the building's chillers and its cooling loads. Most building air conditioning systems operate with chilled water at approximately 44° F.

Chiller efficiency. The ratio of the energy used by a chiller to the cooling produced when the chiller is being operated at design conditions. It is generally expressed in kW/ton. *See also* Coefficient of performance.

Coefficient of performance (COP). The ratio of the cooling produced by a chiller to its energy input when the chiller is being operated at design conditions. See also Chiller efficiency.

Color rendering index (CRI). A measure of the quality of the light produced by a source, a lamp's CRI indicates how well the light source renders colors. Measured on a scale of zero to 100 with 100 being daylight, a CRI of 60 or better is generally considered to be good for most applications.

Color temperature. A measure used to rate the relative color of the light produced by a source. Expressed in degrees Kelvin, the color temperature provides a means for comparing the distribution of light produced by different types of light sources. Lower color temperatures, particularly those below 3,000 K produce more reddish tones while color temperatures above 4,000 K produce bluer colors.

Combustion air. The air supplied to natural gas, propane, oil, and coal fired boilers to support combustion. To ensure complete combustion of the fuel and to limit smoking, the quantity of combustion air supplied is always greater than the minimum needed. However, as the level of combustion air is increased, the combustion efficiency of the boiler decreases. *See also* Excess combustion air.

Condensate. The water that is formed when steam cools and changes from a gas to a liquid in a steam heating or distribution system.

Cooling tower. A unit that is used to reject heat to the atmosphere from air conditioning and refrigeration systems. Air passes by natural or forced flow through the tower, cooling the water by a combination of conduction and evaporation.

Curtailable load. A building energy load that can be turned off when the demand for energy becomes too high. Most frequently associated with electricity use, the switching of curtailable loads can be an effective way to reduce the cost of electrical service. *See also* Demand limiting.

Degree day. The difference between the mean temperature for any day and 65° F, when the mean temperature is below 65° F. Degree days are used to measure the relative severity of heating seasons.

Demand, electrical. Electrical demand, measured in kilowatts, is a measure of the rate at which a facility is using electricity. Electrical utilities charge users a demand charge based on their peak electrical demand for a given month measured over a set interval, typically 15 or 30 minutes. Electrical demand can be a significant component of the overall electrical bill. *See also* Demand limiting.

Demand limiting. The practice of monitoring a facility's demand for electricity and curtailing electrical loads when the demand approaches a predetermined value. Demand limiting is an effective way of reducing the cost of electricity. *See also* Demand, electrical.

Efficiency, boiler. Although there are a number of ways to rate the efficiency of a boiler, the most meaningful measure is a boiler's fuel-to-steam efficiency. Fuel-to-steam efficiency rates the boiler's ability to convert fuel energy into steam or hot water energy. It includes all losses associated with operation of the boiler, including combustion, stack, heat transfer, radiation, and convection losses. A well maintained and operated boiler will operate with a fuel-to-steam efficiency of between 75 and 85 percent.

Efficiency, motor. The ratio of a motor's power output to its input power is its efficiency. Motor efficiency is a direct measure of how effectively a motor converts electrical energy to mechanical energy. A motor's efficiency varies with the load placed on it, generally decreasing with decreasing load. Published motor efficiency ratings are for full-load operating conditions.

Energy use index (EUI). The energy use index is a measure of the relative energy efficiency of a facility. It is calculated by converting all of the energy used by the facility to their Btu equivalents, summing them, and dividing by the total conditioned gross square footage of the facility. It provides a means of comparing energy use in similar facilities.

Energy audit. The energy audit is the systematic evaluation of energy use in a facility to determine how much energy is used where, when, and by what systems. It is an important step in identifying opportunities for energy conservation.

Energy management system (EMS). A microprocessor based system connected to hundreds or thousands of monitoring and control points located throughout a facility. The systems perform a number of functions including energy management, HVAC system controls, equipment monitoring, and fire and life safety.

Excess combustion air. In order to ensure complete combustion of the fuel supplied to a boiler without the development of smoke or soot, the quantity of air supplied to support combustion is greater than what is theoretically required for complete combustion. This additional quantity of air is called excess combustion air. The quantity of excess combustion air required varies with the type of boiler, its condition, and the type of fuel being burned.

Foot-candle (FC). The quantity of light produced by a light source or the quantity of light falling on a surface is measured in foot-candles. One foot-candle is equal to one lumen of light distributed evenly over a one-square-foot surface. It is approximately equal to the intensity of light at a distance of one foot from a standard candle. *See also* Lumen.

Fouling. Fouling is a process that takes place as solids that are in suspension in the circulating water drop out of suspension and accumulate on the surfaces of heat exchangers in boilers, chillers, and cooling towers. The rate at which fouling occurs accelerates with increasing temperature differences and slower water flows. Fouling is controlled through water treatment and regular cleaning of heat transfer surfaces.

Fouling factor. A measure of the thermal resistance across heat transfer surfaces as the result of fouling. New, clean surfaces have a typical fouling factor of 0.0002 or less. As deposits build on the surface, the fouling factor increases, decreasing heat transfer efficiency.

Fuel adjustment charges. Fuel adjustment charges appear in utility bills as a means for utility companies to recover changes in their cost of fuels. When utility rate structures are set, they are based on an assumed cost of fuel to the utility company. Often, particularly when fuel prices are rapidly changing, the estimated average cost is inaccurate. The fuel adjustment charge is used to correct the estimated costs that were used to calculate the rate schedule for the actual fuel costs. Since the original rates were estimated and could be higher or lower than the actual costs, the fuel adjustment charge could be a charge or a credit

Group relamping. The practice of replacing all lamps of a particular type in a given area of a facility at the same time to reduce labor costs and to reduce the effect of lamp lumen depreciation. Most programs base their relamping intervals on the burnout rate of the existing lamps, typically ten to 15 percent. *See also* Spot relamping.

Heat gain. The amount of heat gained by a space from all sources, including conduction, ventilation, the sun, lighting, people, and equipment. In order to maintain comfortable conditions within a facility during the cooling season, the heat gain represents the quantity of energy that must be removed from the facility by the air conditioning system.

Heat loss. The amount of heat lost by a space due to conduction, infiltration, and ventilation during the heating season. It represents the amount of heat that must be added to the space by the heating system in order to maintain comfortable conditions.

Hertz (Hz). The unit measure of frequency in cycles per second. For alternating current in the U.S., the standard line frequency is 60 Hz.

HVAC. Standard abbreviation for heating, ventilation, and air conditioning.

Illuminance. The quantity of light that strikes a surface, usually measured in foot-candles. Also commonly known as lighting level.

Infiltration. Air that leaks into a building through cracks and gaps in the exterior surfaces, particularly around doors and windows. Infiltration contributes to both heating and cooling loads.

Kelvin (K). A unit of measure of temperature measured above absolute zero. Degrees Centigrade can be converted to degrees Kelvin by adding 273. In lighting systems, it is used to measure the color of the light produced by a source.

Kilowatt (kW). A unit of measure for active power in an electrical system equal to 1,000 Watts.

Kilowatt-hour (kWh). A unit of measure for electrical energy determined by multiplying the load in kilowatts by the time it is operated in hours.

Lamp lumen depreciation factor. As lamps age, their light output declines as a result of the deterioration of the lamp's components. The lamp lumen depreciation factor is the percentage of initial lumen output that is produced by a particular lamp after a set period of time. It is used by lighting system designers to account for lower lumen output as the system ages.

Life cycle cost. Life cycle cost is a tool used to evaluate the total cost of owning and operating a system or building component over its expected life. It is particularly useful when evaluating options that have significantly different first costs.

Load factor. The ratio of a facility's average electrical load to its peak load. It is often used as an element in a utility's electrical rate structure.

Load profile. The variation in the electrical energy use for a facility over a set period of time, typically a day, week, month, or year.

Lumen. A measure of light output from a light source.

Make-up water. Water that is supplied to a cooling tower or boiler system to replace water lost through blowdown, leakage, or evaporation.

Mean light output. The light produced by a lamp when it reaches 40 percent of its rated life. In most cases it is significantly less than the initial light output of the source. Mean light output is often called design light output.

Occupancy sensor. A control used in lighting systems to limit the operation of the lighting system to only those periods of time when a person is present in the area.

Occupied hours. The regularly scheduled time when a facility is open for business or occupied. It is used in scheduling the operation of building HVAC and lighting systems.

Payback, simple. The time required for an energy conservation measure to recover its implementation costs through energy savings. It is calculated by dividing the total cost of implementation by the monthly or annual energy savings. Simple payback can also be used to calculate the time required to recover implementation costs for changes in maintenance practices.

Peak shaving. A method for reducing the costs of electrical demand in a facility by temporarily turning off selected electrical loads when the electrical demand approaches a preset level. Loads are typically turned off for five to 15 minutes. *See also* Demand limiting.

Power factor. Power factor is the relationship between active power measured in Watts to the reactive power measured in volt-amps in an alternating current circuit. It is a measure of how efficiently power is being used. Most utilities impose a penalty when a facility's power factor falls below a set value, typically 0.80. *See also* Reactive power.

R-value. A measure of a material's resistance to heat flow expressed in square feet—hour—degree Fahrenheit per Btu. The overall R-value for a building component such as a window is the sum of all of the component R-values. R-value is the reciprocal of the material's thermal conductance. *See also* U-value.

Ratchet clause. Utility companies must install sufficient capacity to meet the peak power demand of their customers, even though that peak may be reached only for a few minutes during one month of the year. To help recover the costs of this required capacity, utility companies install a ratchet clause in their rate schedule. The ratchet clause allows them to charge a customer the peak demand for that month or a percentage of the peak demand for the entire year, typically between 60 and 80 percent.

Reciprocating chiller. A positive displacement compressor that uses one or more pistons to compress a refrigerant gas. Capacity is controlled by regulating the flow of refrigerant gas to the compressor. They are widely used in facilities with cooling loads of 100 tons or less.

Reactive power. Also known as magnetizing power, this is the power required to produce the magnetic flux in induction devices such as motors. Measured in kilovoltamperes (kVa), it contributes to a facility's low power factor. *See also* Power factor.

Rotary chiller. A positive displacement compressor that uses two machined rotors to compress a refrigerant gas between their lobes. Capacity is controlled by regulating the flow of refrigerant to the compressor. Smaller in size than centrifugal chillers, most units have operating capacities between 175 and 750 tons.

Scale. A type of fouling that occurs on heat transfer surfaces in chillers and boilers. It occurs when soluble salts found in the water precipitate out and become attached to heat transfer surfaces. These salts, including calcium sulfate, calcium phosphate, and calcium carbonate form a hard layer that is difficult to remove and decreases the heat transfer efficiency of the chiller or boiler. *See also* Fouling factor.

Service factor, motor. A multiplier that can be applied to a motor's full-load horsepower rating to determine what temporary overload conditions that the motor can operate under. Most motor service factors fall in the range of 1.0 to 1.4.

Shading coefficient. A multiplier used to adjust the quantity of solar gain through clear glass for tinted and reflective glass.

Spot relamping. The practice of replacing lamps one at a time when they burn out. *See also* Group relamping.

Steam trap. A device used to remove condensate, air, and non-condensable gasses from steam systems while preventing the loss of steam.

Time-of-day rates. An metering system used by some electrical utilities that charges different rates for electricity use based on when it was used during the day. There generally are three different time periods and rates; peak, intermediate, and off-peak. Rates can vary by as much as a factor of five between peak and off-peak times.

Tons of refrigeration. The standard unit of measure for cooling capacity. A ton of refrigeration is equal to 12,000 Btu per hour of cooling.

U-value. The heat transfer coefficient of a building component expressed in Btu per hour—square foot—degree Fahrenheit temperature difference. It is the reciprocal of the sum of the all of the individual resistances for composite components. *See also* R-value.

Ventilation air. The portion of the air supplied to an area within a facility that comes from outside the facility. Used to control indoor air quality, ventilation air rates have a significant impact on facility energy use.

Water treatment. The process of adding chemicals to circulating water systems including boiler, chilled, heating, and condenser water systems to control the buildup of scale on heat transfer surfaces. *See also* Scale.

Watt. A unit of measure for electrical power. One Watt equals the product of one Volt and one ampere.

Index

A

Absorption chiller, 54–55, 66–69
Accounting, energy. *See* Energy use
Acoustic testing, 106
Age of facility, and energy use, 33, 35
Age-related factors, in energy use, 33, 35, 152, 153
Air conditioning, 13, 14, 126, 157
  See also Chiller system; Chiller system maintenance
Air handling system maintenance, 109–126
  conservation practices in, 124–126
    day-night setback, 125–126
    economizer cycle, 126
    optimum start-stop timing, 125
    scheduled operation, 124–125
  controls, 118–119
  cost effectiveness of, 110
  dampers, 119
  ductwork, 119–120
  fans, 111–112
  filters, 116—118
  heating/cooling coils, 116
  impact of poor maintenance, 109–110
  implementation of, 120–124
  and indoor air quality (IAQ) problems, 112–113
  and ventilation rates, 112–115
Alignment, motor, 181–182
Alligatoring, 219–220, 228
American Society of Heating, Refrigeration, and Air Conditioning Engineers, 113
Amines, in water treatment, 89
Association of Physical Plant Administrators (APPA), 30
Availability factor, 324, 326, 327

B

Backup system, 291
Ballast, 143, 149
Bearings, lubrication of, 76
Benchmarking, 326, 328–330

Bitumen roof, 218, 231, 235–242
Blowdown, 60, 90–92
Boiler system maintenance, 83–108
  blowdown requirements in, 90–92
  burner, 93, 96
  combustion air controls, 93, 94, 95
  conservation practices in, 107–108
    load management, 107–108
    lower operating pressure, 108
  cost savings through, 84, 85
  efficiency rating for, 84, 86
  operating log in, 86–88
  preventive activities in, 96–100
  water treatment program, 88–90
  See also Steam distribution maintenance
Breakdown frequency, 296, 319–320, 321
Breakdown maintenance, 16, 52, 110, 296, 300
  for doors and windows, 190–191
  overtime hours for, 322, 324, 325
Breakdown torque, 171
Btu. *See* Energy use
Budgets, 20
Building type, and energy use, 29–30, 31
Burner, boiler, 93, 96

C

Carbon dioxide, 24, 93, 95
Carbonic acid, 88
Centrifugal chiller, 53
Chiller system, 51–53, 275–276
  absorption, 54–55, 66–69
  centrifugal, 53
  reciprocating, 53–54
  rotary, 54
Chiller system maintenance, 51–82
  absorption chiller, 66–69
  and backup system, 291
  and corrosion, 60–61, 69
  for energy management, 70–72
    chilled water reset, 71–72
    condenser water temperature reset, 71
    indirect free cooling, 72

Chiller system maintenance *(continued)*
  load allocation, 70
    pilot program, 278
  and fouling, 57–60
  impact of poor maintenance, 52
  implementation of, 79, 82
  leak monitoring, 62, 65
  oil testing, 66
  refrigerant testing, 66
  and scale, 61–62
  scheduled inspections in, 55–57, 62, 63–64
  tube testing, 65–66
  See also Cooling tower maintenance
Clarification, in water treatment, 88
Climate, and energy use, 30, 33, 34, 37, 39
Cogeneration equipment, 4
Colleges and universities, energy use of, 30
Color renderings index (CRI), 147
Color temperature, 145, 147
Commercial Buildings Energy Consumption and
    Expenditures 2012, 276
Computer operations, and energy use, 29–30
Conservation projects
  decline in interest, 7, 16
  and electrical rate structure, 44
  focus on, 6, 7, 8
  vs maintenance activities, 3–4, 6
  maintenance role in, 8–9, 16
    air handling system, 124–126
    boiler system, 107–108
    service hot water system, 135, 137
    See also Energy maintenance program
  payback measurement for, 14–15
  quick fixes, 13–14, 30
Construction factors, in energy use, 33, 36, 37, 38
Contracted energy maintenance program, 297,
    300–304
Contractor, selecting, 303–304, 305
Contracts, maintenance, 302, 304
Controls
  air handler, 118–119
  lighting, 155, 157–159
Cooling coils, air handler, 116
Cooling tower maintenance, 59, 72–79
  centrifugal pumps, 76, 79, 80, 81
  drift eliminators, 74
  fans, 75
  fill material, 73–74
  inspection programs, 76–79
  requirements for, 73
  for structural problems, 75–76
  water distribution system, 74–75

Core sampling, 215–216
Corrosion
  in boiler system, 89
  in chiller system, 60–61, 69
  metal cladding, 258
  metal roof, 243
  in service hot water system, 134
Crystallization, 69
Curtailable service rider, 46
Customer charges, 43

D

Dampers, air handler, 119
Day-night setback, 125–126
Deaeration, in water treatment, 89
Demand charges, 42
Demineralization, in water treatment, 88–89
Doors and windows, 189–207
  and energy losses, 189, 193–198
  glazings, 193, 194, 196, 197
  hardware, 191–192
  impact of poor maintenance, 190–191
  maintenance program, 198–206
    door, 199, 201, 202
    implementation of, 204, 206
    inventory for, 199, 200, 201, 203
    repair *vs* replacement, 206
    window, 201, 204, 205
  overhead doors, 198
  types of construction, 191, 192–193
Drift eliminator, 74
Ductwork, in air handler, 119–120

E

Eddy current testing, 65–66
Efficacy, 145, 146
Efflorescence, 258
Electrical capacitance testing, 216
Electric motor maintenance, 166–186
  alignment check, 181–182
  cost effectiveness of, 166–167
  efficiency considerations for, 167–170
  historical record for, 172, 174
  impact of poor maintenance, 170
  implementation of, 182–183
  inspection program, 180
  insulation testing, 180–181
  inventory data for, 172, 173
  lubrication, 180
  and oversized motors, 177–178

Electric motor maintenance *(continued)*
  and part-load operation, 174, 176, 177
  replacement program, 177–178, 183–184
  shutdown/start-up, 184–185
  vibration measurement, 182
  voltage testing of system, 178, 180
Electric motors
  high-efficiency models, 168–170, 184
  induction-motor operation, 171, 174
  losses in, 167–168
  testing, 174, 175
Electric power
  bill components, 41–43
  generating capacity margin for, 23
  rate structure for, 43–46
  See also Electric motor maintenance; Electric
    motors; Lighting system maintenance
Energy
  demand, 20, 23
  imports, 22
  shortages, 13, 16, 17, 23
  supply, 17, 20–23
  See also Electric power; Natural gas; Oil
Energy audit, 12, 40
Energy budget, 28, 37, 39
Energy Conservation and Production Act of
  1996, 26
Energy Information Administration (EIA), 30,
  33, 37
Energy maintenance program
  contracted program, 297, 300–304, 305
  cost effectiveness of, 9–10, 279, 280, 291
  goal setting for, 287–289
  hybrid program, 304, 306–307
  implementation options for, 293–307
  with in-house personnel, 294–297, 298–299
  inventory of equipment, 289–290
  man-hour requirements in, 289, 297
  measuring effectiveness of, 308–329
    benchmarking in, 326, 328–330
    energy accounting in. *See* Energy use
    performance tracking in, 318–324
  phasing schedule for, 287
  pilot, 277–278
  planning process, 283–292
  selection of equipment/components, 290–292
  selling management on, 273–282
  status reports on, 281–282
  tool and training requirements for, 285, 297
  workforce allocation for, 285–286, 296–297, 306
Energy management
  elements of, 11–12

Energy management *(continued)*
  vs energy conservation, 11
  impact on energy use, 37
  incentives for, 19–25
  maintenance requirements for, 3–10
  oil embargo impact on, 12–13
  performance measurement, 276–277
  phases of, 13–18
  See also Conservation projects; Energy mainte-
    nance program; Maintenance
Energy performance. *See* Performance measurement
Energy Policy Act of 2012 (EPACT), 168
Energy use, 309–318
  per heating degree day, 314, 316
  per square foot, 28, 33, 276–277, 311–312, 313
  per unit of production, 312, 314, 315
  statistical models of, 314, 317–318
Energy use index (EUI), 276–277
Environment, impact of energy use, 23–24
Exterior walls. *See* Foundation and exterior walls

F

Fans
  air handler, 111–112
  cooling tower, 75
Fastener failure, 227–228, 243, 258
Feedback, 12
Felts, in built-up roofs, 218
Filters, air handler, 116–118
Fishmouths, 219
Flashing failure, 220, 228, 235–236, 248
Flow regulators, in service hot water
  system, 137
Fluorescent lamps, 150, 161, 162
Fouling, 57–60
Foundation and exterior walls, 254–269
  causes of problems, 254, 259–261
  impact of poor maintenance, 255
  maintenance program, 261–269
    implementation of, 264, 269
    inspection, 261, 264, 265–268
    inventory, 252–263, 261
    testing, 264
  thermal conductivity of, 254–255, 256, 257
  types of construction, 257–259
Fuel adjustment charges, 43

G

Galvanic corrosion, 61
General service rate, 43–44
Global warming, 24

Goal setting, 287–292
Group relamping, 160–161

H

Hardware, door, 191–192
Heat exchanger tubes, testing, 65–66
Heating coils, air handler, 116
Heating degree day (HDD), 314, 316
Heat and radiation
    in roof problems, 209–210, 219–220, 228
    in wall problems, 259–260
High intensity discharge (HID) light
        sources, 151–152
Hinges, 192
Hot water. *See* Service hot water system
HVAC systems, 13, 14, 112
Hydrogen gas, 69

I

Incandescent lamps, 147, 149
Indoor air quality (IAQ), 112–113
Infrared surveys, 101, 102
Insulation
    air handler, 120
    motor, 180–181, 185
    roof, 208
    service hot water distribution, 134–135
    steam distribution, 100–101
    wall, 258, 260, 264
Ion-exchange, 89

K

Kelvin (K), 145, 147
Kilovars (kvar), 42
Kilovolt amps (kVa), 43, 45
Kilowatt hours (kWh), 41, 44
Kilowatts (kW), 42, 46

L

Lighting system
    efficiency of, 147
    fluorescent, 150, 161, 162
    incandescent, 147, 149
    levels of illumination, 152–155
    mercury vapor, 149–150
    metal halide, 151
    sodium, 151–152
    terminology, 145–147
Lighting system maintenance, 143–165
    automated controls in, 155–159
    cleaning program, 161, 163–164
    goal of, 144
    impact of poor maintenance, 143–144

Lighting system maintenance *(continued)*
    implementation of, 164–165
    inspection program, 164
    relamping program, 159–161
Lime-soda softening, 88–89
Lithium bromide, 69
Load factor, and electrical rates, 43
Load management, boiler, 107–108
Locked-rotor torque, 171
Log
    maintenance, 55, 56, 75
    operating, 86–88, 102
Lubrication, motor, 180
Lumen, 145

M

Magnetizing current, 42
Maintenance
    breakdown, 16, 52, 110, 190–191, 296, 300,
        322, 324, 325
    of building components. *See* Doors and win-
        dows; Foundation and exterior walls; Roofs
    vs conservation projects, 3–4, 6
    of conservation projects, 8–9, 16
    deferred and neglected, 5–6
    lack of understanding about, 7–8
    negative attitude toward, 4–5
    See also Air handling system maintenance;
        Boiler system maintenance; Chiller system
        maintenance; Electric motor maintenance;
        Lighting system maintenance; Service hot
        water system maintenance
Maintenance program. *See* Energy maintenance
    program
Manhole inspection, 101–102
Masonry construction, 257–258
Megohmmeter test, 181
Mercury vapor lamps, 149–150
Metal cladding, 258
Metal doors, 191, 195
Metal framed windows, 193
Metal halide lamps, 151
Meters, 12
Motors. *See* Electric motor maintenance; Electric
    motors

N

National Electrical Manufacturers
    Association (NEMA), 166, 185
Natural gas
    rate structure for, 46–47

Natural gas *(continued)*
  supply and demand, 22–23
Natural gas-fired boilers, 93, 96
Nuclear detection testing, 216

O

Occupancy patterns, and energy use, 30, 32
Occupancy sensors, 158
Oil
  domestic production decline, 17, 21–22
  embargo, 13, 19
  imports, 22
  prices, 16
  reserves, 17, 21–22
Oil-fired boilers, 96
Oil testing, 66
On-off control, 157–159
Optimum start-stop timing, 125
Organization of Petroleum Exporting Countries
  (OPEC), oil embargo of, 13, 19
Overtime hours, for breakdown maintenance,
  322, 324, 325
Oxidation, roof damage from, 210, 219–220

P

Payback, simple, 14–15
Performance measurement
  comparisons to other facilities, 27, 39, 326, 328–
    330
  constant evaluation in, 27
  of energy budget, 28, 37, 39
  factors influencing, 28–39
    building age, 33, 35
    building type, 29–30, 31
    climate, 30, 33, 34, 37, 39
    construction, 33, 36, 37, 38
    occupancy, 30, 32
    people, 29
  standards and codes in, 26
  tracking systems, 318–326
    availability factor, 324, 326, 327
    breakdown frequency, 296, 319–320, 321
    breakdown maintenance overtime, 322, 324,
      325
    maintenance requests, 322, 323
Polarization test, 181
Ponding, roof damage from, 211–212, 227, 235
Power factor, 45
Power-producing current, 42
Pull-up torque, 171

R

Ratchet clause, 42
Reciprocating chiller, 53–54
Reduced voltage starter, 185
Refrigerant testing, 66
Remotely programmed controls, 158–159
Replacement
  door and window, 206
  roof, 252–253
  See also Scheduled replacement
Roofs, 208–253
  causes of problems, 209–213, 218–220, 227–
    228, 235–236
    alligatoring, 219–220, 228
    blistering, 219
    chemical, 210–211
    corrosion, 243
    fastener failure, 227–228, 243
    fishmouths, 219
    flashing failure, 220, 228, 235–236, 248
    foot traffic, 211
    heat/radiation, 209–210, 219–220, 228
    ponding, 211–212, 219, 227, 235
    punctures, 219, 228
    seam failure, 227, 235
    thermal/structural movement, 212–213
    wind, 212, 235
  incidence of problems, 208
  maintenance program, 213–253
    for built-up roofs, 217–222, 225
    cost-effectiveness of, 213–214, 280
    implementation of, 249, 252
    inspection, 214–215, 220, 225, 231, 236, 239–
      241, 243, 246–247, 249, 251
    inventory, 214, 220, 221–224, 228–230, 236,
      237–238, 243, 244–245, 249, 250
    for metal roof, 242–247
    for modified bitumen roofs, 231, 235–242
    repair materials, 225
    repair, recover, replacement decision, 252–253
    for shingle/tile roof, 248–249, 250–251
    for single-ply roof, 223–224, 225–231, 232–
      234
    testing, 215–217, 225, 231, 236, 242
Rotary chiller, 54

S

Scale
  in boiler systems, 88, 90
  in chiller systems, 61–62, 74

Scale *(continued)*
in steam traps, 105
Scheduled operation, 124–125
Scheduled replacement
lamp, 159–161
motor, 177–178, 183–184
Seam failure, roof, 227, 235
Service hot water system
requirements for, 128–131, 135, 136
storage, 131–132
tankless, 132
Service hot water system maintenance, 127–140
conservation practices in, 135, 137
distribution, 134–135
impact of poor maintenance, 127–128
implementation of, 137–138
storage tanks, 133–134
water heaters, 133
Shingle roofs, 248–249, 250–251
Single-ply roofs, 223–224, 225–231, 232–234
Sodium lamps
high-pressure, 151–152
low-pressure, 152
Spot relamping, 159–160
Status reports, 281–282
Steam distribution maintenance
condensate systems, 102–103, 104
piping, 100–102
traps, 103, 105–107
underground lines, 101–102
Structural movement
and roof damage, 213
and wall damage, 260
Surge comparison test, 181

T

Thermal imaging, 216–217
Thermal movement
and roof damage, 212–213
and wall damage, 259–260
Thermographic scans, 231, 236, 242, 264
Thermoplastic single-ply membranes, 227
Thermoset single-ply membranes, 226–227
Tile roofs, 248–249, 250–251
Time clocks, 158
Time-of-day (TOD) rate structure, 44
Torque ratings, 171
Trees and shrubs, wall damage from, 260–261

U

Ultraviolet light, 209–210, 228, 259
Utility bill analysis, 40–47
electrical, 41–46
natural gas, 46–47
U-values, 194, 195–197, 255

V

Variable lighting control, 159
Ventilation air, rate of, 112–115
Vibration, motor, 182
Vinyl framed windows, 193
Volatilization, roof damage from, 210–211
Voltage testing, system, 178, 180

W

Walls. *See* Foundation and exterior walls
Water
alkalinity/acidity, 62, 88
bleed off, 60, 90–92
roof penetration, 211–212, 220, 227, 228, 235, 243, 252
wall penetration, 260, 264
See also Chiller system; Service hot water system
Water treatment
boiler, 88–90
chiller, 59–60
cooling tower, 73
elements in, 88–89
Weather-stripping, 192
Wind
and roof damage, 212, 235, 243
and wall damage, 260
Windows. *See* Doors and windows
Wood doors, 191, 195
Wood frame construction, 257
Wood framed windows, 193
Workforce
allocation of, 285–286, 296–297, 298–299, 306
flexibility, 300
in-house, 294–297, 298–299, 306
man-hour requirements, 289, 297
planning input of, 284–285
training, 285, 298